Chicken

Chicken

The Dangerous Transformation of
America's Favorite Food

STEVE STRIFFLER

Yale University Press   New Haven and London

Published with assistance from the Kingsley Trust Association
Publication Fund established by the Scroll and Key Society of Yale
College.

Set in Minion Roman type by Integrated Publishing Solutions.
Printed in the United States of America by Vail-Ballou Press.

The Library of Congress has cataloged the hardcover edition as
follows:

Striffler, Steve, 1967–
Chicken : the dangerous transformation of America's favorite food
/ Steve Striffler.
        p.   cm.—(Yale agrarian studies series)
Includes bibliographical references and index.
ISBN 0-300-09529-5 (hardcover : alk. paper)

1. Chicken industry—United States—History. 2. Poultry plants—
United States—History. 3. McDonald's Corporation. I. Title. II.
Yale agrarian studies.
HD9437.U62S77 2005
338.1'765'00973—dc22   2005007310

ISBN 978-0-300-12367-8 (pbk. : alk. paper)

A catalogue record for this book is available from the
British Library.

10 9 8 7

Contents

Preface

It is my first day of work in one of the largest poultry-processing plants in the world. I am given "the tour" that all new workers receive. We begin in live hanging. Hundreds of live chickens flood off the trucks, down a chute, and into a bin where workers quickly hang them by their feet onto the production line. It's surreal. It is nearly pitch black, on the theory that the darkness soothes the terrified birds. The smell and look of the place are oppressive, so I look for something to focus on other than the hanging itself. A worker. I eventually learn that Javier is from Mexico, but the figure is hard to make out at first. He is covered from head to toe in protective clothing that is itself coated with blood, shit, and feathers. Javier's job is simple, if somewhat gruesome. The chickens have already passed through scalding hot water and have been electrocuted, a process designed to both kill the bird and begin the cleaning. Neither task is accomplished perfectly. The communal baths, popularly known as fecal soups, do clean, but they also pass harmful microbes from one bird to the next. The bath also doesn't do a particularly good job of killing the chickens: one out of every twenty seems to make it through alive. The birds are in their last stages of life when they reach Javier. For eight hours a day he sits on a stool, knife in hand, and stabs the few chickens that have managed to hold onto life.

While watching Javier, I realize what this book will be about. How did Javier and the chickens arrive in this place, under these conditions? Where do they go once they leave the plant? And what does their experience in the plant mean to those of us who eat chicken? The search for answers led me to study a period when chickens were raised and processed quite differently, and to visit poultry farms, supermarkets, restaurants, and communities in the southern United States and central Mexico. As I learned while doing this research, whereas the chicken's journey is one characterized by uniformity and predictability, the worker's path is defined by variation, insecurity, and chaos. Neither experience leads to a particularly healthy outcome for bird, worker, farmer, environment, or consumer.

I do not feel sorry for Javier or the chickens. I have worked in a plant before, and stabbing chickens is a relatively easy job. Many workers would be glad to trade places. And the chickens are there to die. I knew this going in. The problem, which became more transparent as I passed by "evisceration," the "KFC line," and the "wing room," was that no one departed from the plant in particularly good shape. The workers left poor, exhausted, and, in many cases, seriously injured. The chickens not only exited the plant dead, but in a "further-processed" form that was not particularly healthy for consumers. In short, the postwar promise of the industrial chicken—as a healthy, plentiful alternative to beef—has been lost for all of the people involved in its raising, processing, and consumption. There has to be a better way.

Acknowledgments

I used to eat chicken without much thought about where it came from, or how and by whom it was raised and processed. Life was much easier then. For complicating my dinner, giving me an education, and allowing me to write this book, I thank a number of people. I begin with everyone involved in the industry, including workers, farmers, managers, and many others, who took the time to share their stories and knowledge. I have the greatest respect for the work you do. Libraries and librarians were also a key part of this long process, particularly those associated with Wilkes Community College, Wilkes County Public Library, Shiloh Museum, and the University of Arkansas. The year I spent as a Rockefeller Fellow at the University Center for International Studies at the University of North Carolina at Chapel Hill gave me precious time to read, write, and think in a stimulating atmosphere. Many thanks to Don Nonini, James Peacock, Niklaus Steiner, Ajantha Subramanian, Harry Watson, and many others for making this possible. Leon Fink's encouragement, as well as his own work, has been crucial to this project. Bill Roseberry, who died before a page of this book was written, remains a guiding force. His advice and assistance during the project's early stages were inspirational.

James Scott was an important influence throughout this project's development. Jim, Mary Summers, and a number of other scholars at Yale University had the great idea of hosting a truly wonderful conference: "The Chicken: Its Biological, Social, Cultural, and Industrial History." Jim and the Yale Program in Agrarian Studies then invited me to present some of my research at what remains one of the most stimulating seminars in academia. I benefited immensely. Finally, Jim also put me in touch with Jean Thomson Black and Yale University Press. Jean is an editor who has not only great vision, but also a unique ability to help authors expand and realize their own ideas. Julie Carlson, the manuscript editor, did a wonderful job tightening and clarifying the book's presentation and context. Thanks to Jean and Julie, Laura Davulis, Jeffrey Schier, the anonymous reviewers, and the press for all the hard work.

A version of Chapter 6 appeared in *Labor History* (August 2002; http://www.tandf.co.uk/journals/titles/0023656X.asp). Thanks to the journal for permission to reprint.

I also thank my parents, Chuck and Nancy, for their generous support. Karon, Tara, Brad, Allie, Oliver, Wallace, Sophie, Elliot, and Pomona all made it as difficult as possible to write this book. May it always be so.

Introduction

F ood and immigration. Both conjure up powerful images and feelings. Both are central to how we understand ourselves as a nation and as a people. And both are topics that most Americans feel profoundly ambivalent and guilty about.

The importance of food is obvious. We are what we eat. For much of our history the family farm anchored our economy, society, political system, and sense of ourselves as Americans. Along the way the family farm gave way to agribusiness, the most productive system of growing, delivering, preparing, and consuming food the world has ever known. Our eating habits, appearance, and health have all changed dramatically as a result of this revolutionary method of delivering food.

Indeed, the very abundance of this system has contributed to a peculiarly modern food crisis. As government agencies, nutrition groups, self-help guides, and the news media constantly remind us, we eat too much overall, especially of the least healthful foods. According to a report on *ABC World News Tonight with Peter Jennings,* for example, American farmers produce nearly twice as much food as America needs.[1] Nearly two out of every three Americans (130 million people) are overweight, and almost one-third (60 million) are obese, an epidemic that has implications for heart disease, cancer, stress, diabetes, and a wide range of other serious health problems.[2] In fact, the Centers for Disease Control and Prevention warns that obesity is rapidly catching up to smoking as the number one cause of death in the United States. Tobacco use killed 430,000 people in 2000, or 18.1 percent of those who died, while poor diet and physical inactivity led to 400,000 deaths in the same year (16.6 percent), up from 300,000 in 1990.[3] Health and Human Services Secretary Tommy Thompson put it bluntly: "Americans need to understand that overweight and obesity are literally killing us."[4]

Today's youth—tomorrow's adults—are also heavier than ever before. There are two times as many overweight children and three times as many overweight adolescents as there were twenty-five years ago.[5] Fifteen percent of school-age children are dangerously overweight, and kids are eating up to two hundred more calories a day than they did just fifteen years ago.[6] In New York City, 43 percent of elementary school kids are overweight or obese.[7]

The information is out there. We hear over and over again that we are fat, that our relationship to food is unhealthy. More importantly, we are becoming aware that our food choices are not entirely innocent. They are determined at an early age and to a large degree by a food industry that is concerned first and foremost with profit.[8] And because unhealthy, processed foods are simply more profitable than healthy ones, the food industry spends most of its annual $33-billion advertising budget "to promote the most highly processed, elaborately packaged, and fast foods. Nearly 70% of food advertising is for convenience foods, candy and snacks, alcoholic beverages, soft drinks, and desserts, whereas just 2.2% is for fruits, vegetables, grains or beans."[9] In 2003, more than 2,800 new candies, desserts, ice creams, and snacks hit the shelves, compared with only 230 new fruit or vegetable products.[10] The average American child views ten thousand food advertisements a year on television alone, and most of these are for fast food, sugared cereals, soda, candy, and other foods filled with fat and calories. This advertising has helped generate a profound cultural shift. Junk food has become what children expect to eat.[11]

Unfortunately, the food industry has help in leading us to the wrong choices. According to the ABC program mentioned earlier, "The processed food industry and the government know what is happening—and they are making it worse."[12] Nutrition expert Marion Nestle agrees that the government promotes overeating. The federal government pumps $20 billion a year into agriculture without thinking about its implications for consumer health. As a result, much of what gets produced with government subsidies is not particularly healthy. Since the mid-1990s, for example, meat and dairy received about three times the subsidies of grains, while fats and oils collected twenty times more government handouts than fruits and vegetables.[13] The government, in short, supports the production of food, and the food industry, without considering how this support affects the health of the American public.

The irony is that despite all the public attention given to the "food crisis," as well as our own personal wars with food and fat, we know surprisingly little about where food comes from or how it is grown and processed. Few of us produce our own food, and most of what we eat now travels great distances and is transformed during the journey. We sense that the process of food production has been corrupted, and we pore over labels to determine fat, sugar, and carbohydrate content, but the origins of the foods, as well as those strange ingredients we can't pronounce, are largely unknown to us.[14] Even with the most basic of foods, few of us really know what we are purchasing and eating.

The importance of immigration for our national psyche is also obvious. We take great pride in our immigrant history. Americans believe that our tolerance and willingness to embrace people from all over the world have made this country great. But we also know that our nation's history is far more complex than a school textbook version can capture. Africans were forced to "immigrate" in one of the great tragedies of human history. Millions more dreamed the American Dream only to encounter hostility from those who had arrived on Ellis Island just before them. Similarly, today those who fail to "blend in" are somehow flawed. Immigrants are simultaneously embraced and rejected. The politicians who most vocally oppose immigration, and ride anti-immigrant sentiment to political office, are often the very same leaders who facilitate the entry of immigrant workers on behalf of corporations that require cheap labor.

We hate food. We love food. We fear immigrants. We need immigrants.

From our inception as a nation, when the first immigrants landed on Plymouth Rock and shared that famous turkey dinner with the natives, food and immigration have been inextricably linked. America's fields, canneries, and processing plants have been rites of passage for thousands of immigrants from Europe, Asia, and Africa. Agricultural labor has been a stepping-stone that has often allowed first-generation immigrants to acquire their own farms or find factory work in America's burgeoning cities.

Mexican immigration in particular has been about ensuring a steady supply of cheap food. By the early part of the twentieth century, California, which was rapidly becoming the most productive agricultural region in the world, was home to thousands of Mexican immigrants who worked the fields and sustained the state's economy. Today, most of the labor of producing and processing food in the United States is done by Latinos, a majority of whom are immigrants from Mexico and, increasingly, Central America.

Much of the behind-the-scenes labor that goes into preparing, serving, and cleaning up food is also done by Latin Americans. An American teenager may take the order, but a Mexican is likely to be cooking the McNuggets and cleaning the deep fryer. Latinos are becoming virtually synonymous with food preparation and cleanup in our nation's restaurants. To find a meal that has not at some point passed through the hands of Mexican immigrants is a difficult task.

And yet few of us think of food as having a history. It just appears, magically disconnected from the people who produce, process, and serve it. Our modern food system encourages us to ignore the connections between what goes into our mouths, how it gets there, and who produces it. Until recently, this willful ignorance has meant little in terms of consumer health. As consumers, we simply benefited—in the form of affordable and abundant food—from the cheap labor of others. The problem is that we now have a food system that not only is dependent on cheap labor, but also requires an easily exploitable workforce to produce and process unhealthy foods. Americans are destroying their bodies by consuming the wrong foods, and immigrants are destroying their bodies by producing those foods. In addition, the system is taking a toll on the general environment as well as on the species of animals and plants we consume. Seen in this light, the cost of food is astronomical.

Overprocessing has revolutionized food in America. The goal of food conglomerates, a handful of which now control most of the world's food supply, is not simply to produce more food. It is to add "value" to food, either by enhancing existing foods or by creating entirely new products, to make more profit. Is processing inherently bad? Of course not. Many foods have to be processed to be edible. Processing also increases variety in our diet, preserves food, and makes our lives more convenient. But do we really need more than 250 different types of breakfast cereals, 450 kinds of sodas, and thousands of different chicken products?[15]

Indeed, recently it seems as though the more we do to a food, the less healthy it becomes for consumers. (Perhaps this is because the easiest and cheapest things for companies to add to food—sugar and fat—also happen to be the least healthy to consume.) Overprocessing is also unhealthy for farmers, who must conform to more exacting standards while losing an increasing share of the food dollar to processors and marketers. And it is unhealthy for workers, who do increasingly tedious, repetitive, dangerous, low-paid, and, in some cases, unnecessary forms of labor in order to create food.

Why Chicken?

The central tensions and contradictions that characterize the American food system are brought into sharp relief through a study of chicken. These include the rise of obesity as America's leading health problem; the influx of immigrants and the Latinization of the low-wage labor force; the concentration of corporate power; the industrialization of cuisine; and the demise of the family farm. Early on, the chicken industry was seen as a way for marginal farmers to supplement their income. Raising chickens would help the rural poor survive the lean times. For consumers, too, the greater availability of chicken was a sign of prosperity. Herbert Hoover's 1928 promise to put a "chicken in every pot" resonated with American voters. Later on, long after most consumers actually did have a chicken in the pot, the bird gained even greater popularity as a healthy alternative to red meat. Eating chicken, so we thought, would provide us with an important source of protein that was relatively free of fat and cholesterol.

American farmers did produce massive quantities of chicken in the postwar period, far more than anyone might have imagined in 1928. Chicken did become available to consumers to a greater degree than Herbert Hoover probably envisioned when he uttered his famous slogan. The rise of chicken—its production, processing, and consumption—has been nothing short of astounding. But even as chicken provided marginal farmers with extra income, it has also ensured that they would remain poor and, after 1960, thoroughly dominated by large corporations that owned or controlled every facet of the industry.

For consumers, the healthful promise of chicken was hijacked before it could be fulfilled. One of chicken's greatest ironies, and one it shares with many of the foods we eat, is that even though we put it on our dinner plates in the 1970s and 1980s because of its healthful qualities, it has subsequently become part of the food problem. Chicken became cheaper and more available. It has also, since the 1980s, come to us in increasingly unhealthy forms. Boneless and skinless breasts notwithstanding, Americans have come to know chicken in the form of nuggets, fingers, strips, and wings. And along the way we have gotten a lot fatter.

The chicken industry has disappointed workers as well. It has provided an important source of employment for the most marginal workers among us. But jobs in processing plants have long been among the lowest paid, least pleasant, and most dangerous in America. For example, jobs at Tyson Foods, the industry leader, are so dangerous, strenuous, and low paying that the turnover is around 75 percent annually. Further, in 1999 Tyson was named one of the "10 Worst Corporations of the Year" by the *Corporate Crime Reporter*, because of seven worker deaths, fines from the Occupational Safety and Health Administration (OSHA), and other violations including child labor (see Chapter 8).[16]

The rewards for working in this industry are certainly meager. Even as industry profits between 1980 and 2000 more than tripled, the real wages of approximately 250,000 poultry workers remained largely stagnant.[17] And the work in the factories has become more intense, in part because of the growing importance of processed chicken products, which has resulted in more tedious, repetitive work and faster production lines. To fill these undesirable jobs, the poultry industry has sought a vulnerable labor force that works more for less and that is hampered in its efforts to unionize.

How do we raise, process, engineer, transport, market, and eat chicken, and what are the connections between these processes? How have forms of farming, work, product development, and consumption changed over time? To understand the ways in which people propelled (or were caught up in) these endeavors, I tell this history through numerous smaller stories about the farmers, workers, scientists, truckers, marketers, corporations, and consumers who are involved with industrial chicken.[18] To do so, I draw on a wide range of sources, including scholarly studies, government reports, media accounts, corporate publications, interviews, and firsthand observation.

Chapter 1 is a brief popular history of chicken from the perspective of consumption. How is it that we came to eat chicken the way we do? The focus is on changes in the way we eat, which, in the case of chicken, is ultimately a story about the rise of processed poultry products. The chicken's postwar success with consumers was due in large part to its relative affordability and healthiness when compared with other sources of animal protein. How, given this promise, did we ultimately end up with a processed food that is not particularly cheap or healthy? Although we as consumers certainly share some of the blame, the development and clever marketing of thousands of new poultry products led consumers to the wrong choices.

Chapter 2 charts the development of the poultry industry during the decades before and after World War II. Emerging first in the Delmarva Peninsula of the eastern shore of Delaware, Maryland, and Virginia, large-scale production of chicken finally became centered in the southern United States. This geographic shift was shadowed by an almost simultaneous transformation in which farmers and small businessmen lost control over chicken production to large agribusiness.

Chapters 3 and 4 continue the story on the local level. Wilkesboro, North Carolina, emerged during the 1940s and 1950s as an industry hub and home of Holly Farms, one of the largest poultry companies in the world by the 1980s. Chapter 3 explores the rise and fall of Holly Farms in particular, and the broader process of industry consolidation more generally. As we will see, Holly's corporate life was defined by mergers and acquisitions; so too was its death. In the late 1980s, Arkansas-based Tyson Foods acquired the Wilkesboro giant in one of the nastiest takeovers in corporate history. Such mergers have defined the food industry in the postwar period and have had profound effects on how we grow, process, and eat chicken. Chapter 4 explores some of the consequences of these mergers for workers and farmers.

Part II explores chicken's growing dependence on an immigrant labor force. By transforming poultry-processing plants and other workplaces, immigrants have also changed much of America's heartland in the South and Midwest during the past quarter century. The influence of industrial poultry extends into our communities, schools, and churches. Chapter 5 focuses on a legal case—well covered in the media—in which Tyson Foods was indicted by the U.S. Immigration and Naturalization Service (INS) for smuggling illegal immigrants into the country in order to work in its processing plants. This admittedly extreme example is perhaps most significant because it highlights the complexity, and hypocrisy, of the relationship between food and immigration, suggesting that chicken depends on often exploitative sets of social relations. It also demonstrates how much time and resources immigrants spend in order to obtain jobs that most Americans do not want.

Chapter 6 is an account of my work in a poultry-processing plant. This is not the first time an "outsider" has worked in a factory; similar work has been done within the meat industry by both scholars and journalists.[19] Nevertheless, I learned a lot about what it takes to produce food by working "on the line" for two summers. I hope to convey some of these lessons or, at the very least, give a sense of the pressure, intensity, pain, and pride that workers experience when turning live birds into some of Americans' favorite foods.

Chapter 7 looks at how Latin Americans, and poultry workers in particular, have been received by communities in America's heartland. The food industry, and especially meat processing, has altered the racial–ethnic makeup of Middle America. I traveled to Siler City, North Carolina, a "chicken town," in order to highlight the problems and opportunities that immigration poses for both "us" and "them."

Chapter 8 points to an alternative way of doing chicken that is more healthy for workers, farmers, consumers, the environment, and the chickens themselves. The positive example of Friendly Chicken suggests that when American food workers and farmers are empowered and adequately compensated for their work, and when care is taken in the handling of chicken, a healthy food can be made for not much more money than the current choices.

I
A New Bird

I

Love That Chicken!

Chicken, an afterthought on American farms before World War II, has been transformed into the most studied and industrialized animal in the world.[1] It has gone from one of the most expensive and least desirable meats to an affordable source of protein that most Americans today consume frequently and with unthinking regularity.[2] Herbert Hoover's 1928 campaign promise seemed like political hyperbole at the time, but for most of the post-war period there has been a "chicken in every pot." And when health-conscious consumers began to turn away from red meat in the 1970s, white-meat producers were ready for the increase in demand.

During the decades surrounding World War II, American farmers, workers, scientists, feed dealers, and many others succeeded in drastically reducing the cost of chicken, delivering untold numbers of chickens to consumers, improving the health of Americans, and making a little money along the way. To be sure, there was a less glorious side to this capitalist miracle even in the early days of the industry. Like the food industry as a whole, power within poultry became concentrated in the hands of a few corporations. Poultry "integrators" came to control nearly every link of the chain, from egg production and the delivery of chicks to the processing and marketing of broilers. This concentration of power came at the expense of both chicken farmers, who lost control over their independent operations, and processing plant workers, who worked hard and earned little (see Chapter 2).

Nevertheless, the agroscientific revolution that transformed much of the food industry led to astounding gains for purveyors of chicken. The amount of time required to turn a day-old chick into a full-grown broiler decreased by almost 20 percent between 1947 and 1951 alone. At the same time, the bird required less feed. In 1940, chickens required more than four pounds of feed for every pound of weight gained. By the late 1980s, this figure was down to around two pounds. As a result, from the 1920s to the mid-1950s the price of chicken declined steadily (and faster than the prices of its main competitors). In the late 1980s, the real price of chicken was less than one-third of its cost in 1955.[3]

Moreover, until very recently, Americans were consuming most of their chicken in very healthy forms: the whole bird, or the whole bird cut into its basic parts. Herbert Hoover's slogan was a promise of economic prosperity that referred to a form of chicken and method of preparation familiar to Americans at that time. For the next fifty years, until processed chicken arrived on the American consumer's radar, chicken became increasingly less expensive and more readily available, and it was consumed in these healthy forms. There was no downside for consumers.

How quickly things change. For today's generation, cooking a whole chicken in a pot is almost unimaginable. A "chicken nugget in every fryer" is a more apt slogan for those who have come of age in the past quarter century. Indeed, McDonald's Chicken McNugget is a particularly useful historical marker. At the time of its introduction in the fall of 1983, Americans were still largely eating chicken in its most basic form. Processing was barely a shadow of its future self and was largely restricted to freezing or cutting the bird into parts. Butchers were becoming part of our nostalgic past, but the range of poultry products was still quite limited by today's standards.

The Chicken McNugget instantly made McDonald's, already the largest user of beef, the second largest user of chicken (behind Kentucky Fried Chicken). During the twelve-week period in which McDonald's introduced the McNugget, the company used five million pounds of chicken per week. McDonald's was expected to sell some 3.5 billion McNuggets during 1984 in its more than six thousand restaurants.[4] More importantly, the McNugget became the cornerstone of the chain's appeal to kids. For the next two decades, children—the McDonald's customers who bring so many others through the golden arches—came for McNuggets as the company expanded to around thirty thousand restaurants in over one hundred countries.[5]

McDonald's did not introduce Americans to chicken. In fact, its introduction of the McNugget demonstrated that, by the 1980s, chicken was already an everyday food—and one that was becoming more popular than red meat. In 1984, Americans were eating twice as much chicken as they had been only two decades before. Meanwhile, beginning in 1976, beef consumption had begun a slow but steady decline. The expense of beef, combined with growing health concerns and questions about its safety, had begun to take a toll.[6]

Statistics tell part of the story. Between 1976 and 1989, per capita chicken consumption rose 50 percent, pushing chicken

ahead of beef as America's favorite meat.[7] By the end of this period, one in three Americans claimed to be eating more
chicken and less beef, due in large part to concerns about fat
and cholesterol.[8] To put it another way, by 1980, after several
decades of growth, the average American ate nearly thirty-
three pounds of chicken each year. But by 1995, only fifteen
years later, the amount consumed increased to more than fifty
pounds, and by the millennium that figure had climbed to
roughly eighty pounds—while per capita consumption of beef
fell to under seventy pounds, and pork consumption hovered
near fifty.[9] By 2001, the typical American household was serving chicken seven times a month, with the average person consuming an amazing eighty-two pounds of chicken a year, or
three times the amount eaten in the 1960s.[10]

Chicken became king. But its ascent during the 1980s and
1990s was very different from its initial rise during the decades
surrounding World War II. Beginning in the 1980s, or roughly
around the time McDonald's introduced the McNugget, the
consumption of *processed* chicken rose remarkably and accounts
for much of chicken's success during the past two decades.
In this respect, the McDonald's McNugget not only reflected
chicken's postwar success, but also anticipated its future form.

To be sure, McDonald's did not introduce us to processed
chicken any more than it introduced us to the bird itself. By the
mid-1980s, we were eating chicken in foods such as franks,
sausages, patties, baloney, loaves, and pastrami. Chicken could
be bought in supermarkets, delis, fast-food joints, convenience
stores, and almost any restaurant. Consumers could buy the bird
whole or in parts, fresh or frozen, bone-in or deboned. Nevertheless, when the McNugget was introduced, most chicken was
still consumed in a relatively nonprocessed form. Further-
processed chicken (anything other than the whole chicken or

cut-up chicken parts) accounted for a mere 16 percent of total consumption. This figure was up from almost nothing only a decade before, but it meant that over 80 percent of chicken was still consumed in a form that resembled the real thing.[11] During the two decades after McDonald's introduced the McNugget, those percentages were reversed. Further-processed chicken *became* chicken, especially for those under the age of thirty. As an undergraduate once confessed to me: "I can't imagine a time before chicken nuggets and strips. That's the only way I eat chicken." It happened that fast. Anyone over thirty-five remembers a time when there was virtually no processed chicken. Those younger than age thirty, however, often forget that chicken comes in any other form.

It is important to keep in mind, then, that the success of chicken and the success of the McNugget, or further-processed chicken in general, are two distinct phenomena. The early rise of chicken was driven by a combination of, on the one hand, consumer demand for a cheap, healthy source of animal protein and, on the other, an agroscientific revolution carried out by farmers, workers, scientists, and others that delivered large quantities of affordable chicken to consumers by the end of the 1950s. By contrast, the more recent rise of processed chicken has been only partially about consumer demand for "convenience foods."

Engineering and Marketing Chicken

Processed chicken is a product of food engineering and clever marketing, designed to combat a problem that has plagued the industry from its inception: Chicken in its most basic form is simply not that profitable. Historically, profit margins within the meat industry have been not only slim, but also unstable

Chicken Today (Photo by Kelley O'Callaghan)

or, at best, cyclical. The key costs in producing chicken are feed and labor. Labor costs are fairly constant and can only be squeezed so much. Feed, however, is made up primarily of commodities (corn and soy) whose prices tend to boom and bust. With already low profit margins, this unpredictability in feed price both made life difficult in the industry and facilitated its consolidation. The bigger firms could weather the lean times, cranking out unprofitable chickens in order to hold onto market share and wait for the good times to return. The smaller firms went out of business or were taken over by the larger enterprises.[12]

But profits generated from industry consolidation and increased efficiency were limited, decreased over time, and could never resolve the basic problems of wild price swings and low profit margins. As a result, some within the industry realized that it was not enough to just produce more chicken: they had to do more to the bird. By adding a little "value" to the standard product, they were able to distance their product from the basic commodity from which it originated.[13]

Enter food engineering and marketing. The first steps in creating new chicken products really began only about fifteen years before McDonald's trotted out the McNugget. In the late 1960s, North Carolina–based Holly Farms, a regional poultry producer that supplied the southeastern United States, began selling prepackaged chicken parts, a step that not only allowed supermarkets to bypass the butcher, but also increased profit margins and enabled companies like Holly to market chickens in a variety of ways.[14] In the early days, such advances were not part of a coherent strategy to develop "value-added products" as much as they were haphazard attempts to introduce more profitable product lines. When Tyson, for example, introduced the Rock Cornish game hen in the 1960s, the idea was simply to produce something a "little special" that would fetch a slightly higher price. Don Tyson quickly learned, however, that if a new product line caught on, he could quickly dominate the market and make real money.[15]

In the early 1970s, Tyson Foods began selling chicken in seven different forms and by the end of the decade had some two dozen products. This surge would be dwarfed in the 1980s and 1990s when the company introduced *thousands* of different poultry products. As Tyson noted in 1979, "We are moving away from being a commodity company toward being a marketing company with specialized products using the Tyson brand." Why sell a plain old chicken when you can offer frozen chicken Kiev dinners?[16] By the mid-1980s, an executive from a major supermarket chain could correctly (and without irony) claim that Tyson was "light-years ahead of the industry in taking chicken and giving you another product out of it."[17] By removing bones, breading chicken parts, or churning out bite-sized chunks, Tyson and other poultry companies were able to charge much more than they could for the ordinary bird.[18]

The first nugget was invented in the early 1970s. Several years later Tyson introduced a mass-marketed chicken-breast patty (a decade later the company would boast twenty-six different kinds).[19] Burger King sold a fast-food chicken sandwich as early as 1977, and then the McNugget solidified the place of processed chicken in American culture.[20] Invention followed invention until the chicken arrived in every conceivable form, flavor, texture, and venue. Further processing also allowed poultry companies to use up those portions of the chicken that Americans, who were increasingly breast crazy, did not want. Through processing and ingenuity, unprofitable by-products such as skin, necks, and backs found a home in hot dogs, nuggets, pet food, and other inventions.

The products themselves, especially in the early years, were often developed by folks like Robert Baker. During the twenty-five years between 1960 and the introduction of the McNugget, Baker created over fifty poultry products, including chicken baloney, steak, salami, chili, hash, pastrami, and ham. He was also chair of Cornell University's poultry science department, which routinely developed and test-marketed products for companies that then adopted them at no cost. Most of the products—including the chicken hot dog, which quickly captured 20 percent of the hot dog market—wound up in supermarkets. As Baker explained in 1984: "Chicken . . . was a loss leader pretty much from 1955 to 1970. So the industry was in bad shape. Once we started the convenience industry, put the chicken in different forms, consumption started going up again."[21] We might add that what consumers were eating was no longer the chicken of Herbert Hoover, but that of university laboratories and fast-food restaurants.

Advances in chicken engineering were uneven during the 1960s and 1970s. Even with the help of universities, poultry companies had to make a considerable investment in order to create a new product (which often required advances in production technologies) and market it. Further, consumer acceptance of new processed chicken products was far from a sure thing. Chicken was hot, but there was a limit. Tyson, for example, introduced a "giblet burger" in order to use up the millions of gizzards it processed. The burger flopped in California stores, and the Arkansas prison system actually refused to inflict the product on inmates.[22]

Despite the occasional misstep, however, the push toward further processing was unrelenting. Poultry companies had little choice. Simply selling more chicken was not a viable option, especially in the long term. Companies like Tyson and Perdue had to keep offering more novel chicken products.

Branding Chicken

Accompanying the continual creation of "new" chicken products was the development of branded chicken. Before the invention of further-processed chicken, producers had little money to spend on advertising; they therefore left marketing to retailers, who, in turn, treated chicken as a loss leader, a product whose only value was bringing customers through the door. Then came Frank Perdue, founder of Delmarva-based Perdue Foods. In 1968 he started using major media, including radio and TV commercials as well as print ads and outdoor messages, in various northeastern cities. His goal: to establish a brand name that was associated with quality. The idea was not only to increase demand for his product (as opposed to generic chicken), but also to enable him to sell Perdue Chicken at a premium price. Perdue's "superior" bird would be sold right next to the brandless supermarket varieties.[23]

Perdue's plan worked. Between 1968, when he began his advertising campaign, and 1974, sales at Perdue more than quadrupled. New York City was the target market, in part because New Yorkers have a reputation for being demanding consumers who recognize (and will pay for) quality. When the advertising campaign began in 1968, Perdue was selling 50,000 broilers a week in New York, or about 1 percent of the total market. A year and a half later, Perdue was selling 350,000 birds a week and would soon control over 25 percent of the market.[24]

Perdue's initial campaign targeted women. According to Ray Kremer, an advertising executive who handled the Perdue account during those heady days of expansion, TV became an important vehicle in 1969: "We wanted to show housewives exactly what the original red, white, and blue Perdue wing tag looked like so they could make sure they were getting one of the company's chickens. [We] placed 60 or 70 ten-second [TV] spots weekly, buying vehicles with substantial women's audiences. Buys included afternoon quiz shows on ABC, spot participation on NBC's *Today Show,* [and] CBS soap operas." Perdue even bought spots during NBC's coverage of the *Apollo 11* splashdown. A cartoon of a chicken wearing a space helmet was accompanied by the headline "Perdue Chickens Are Out of This World."[25] Clever stuff.

After considerable success, however, the gains from the advertising campaign began to taper off. Consumer awareness of the Perdue brand stalled at just over 20 percent. Perdue needed not only to increase awareness of its brand, but also to give consumers a reason to pay more for its chicken. The answer to the dilemma turned out to be crotchety Frank Perdue himself. Cartoons of chickens were replaced with Frank Perdue attesting to the quality of his chickens as he inspected Perdue plants or watched one of his delivery trucks speed off to New York. Perdue was not known for being particularly lovable, but the ad campaigns turned his abrasive nature into a down-home, no-nonsense sort of charm: as his famous phrase put it, "It takes a tough man to make a tender chicken." He even targeted the Hispanic community, appearing on Spanish-language TV with a simple message: "My chickens are yellow. I'm the Chiquita of chicken."[26]

Perdue's success was phenomenal and not lost on the company's competitors. Despite the received wisdom that there was no need to advertise chicken because of soaring demand, branding swept the industry during the 1970s. As a result, by the mid-1970s Americans had abandoned their long-standing practice of selecting chickens based on plumpness, color, and faith in local supermarkets, and instead were jumping on the chicken brand-wagon.[27] Radio advertisements aimed at hungry customers driving home from work.[28] Pearl Bailey told Americans to put a "Paramount in every pot" and reassured us that Paramount Chickens (a subsidiary of Cargill, Inc.) looked after its birds "just like a mother hen."[29] Dinah Shore reminded consumers that "America's Cookin' with Holly Farms."[30] In Denver, Tyson began a brand-name test campaign at the beginning of the decade and quickly became an industry leader.[31] Country-western singer Tom T. Hall sung the praises of Tyson chicken before Mary Lou Retton told us what kind of chicken America's favorite gymnast ate.[32] And by 1989, Tyson was spending $14 million on a TV campaign to announce that it was "Feeding You Like Family."[33]

The importance of branding increased as the number of chicken products proliferated during the 1980s. By 1980, about one-third of all chicken sold in supermarkets was branded. Five years later almost 50 percent of chicken bought in supermarkets carried a brand name, a fact that helps explain poultry's higher profit margins when compared to beef (which remains relatively unbranded even today). According to a *Consumer Reports* survey, consumers were more likely to buy branded chicken largely because of perceived differences in quality.[34] Yet, as the study also found, paying more for a brand did little to guarantee superior taste or more consistent quality. Through the magical world of advertising you could get consumers not only to buy a particular brand of chicken, but actually to pay premium prices for the same product. Branding added "value" to chicken without adding any real value to chicken.

Fast Chicken

By 1980, then, the groundwork for processed chicken parts had been laid. The McNugget was both a historical marker and a force to be reckoned with. As the *New York Times* noted, "What really ignited the further-processed stampede was McDonald's introduction and vigorous promotion of Chicken McNuggets." Within weeks, McDonald's was one the largest users of chicken in the world, purchasing around 1.5 percent of the nation's total broiler production for its McNuggets.[35] In less than two years, the McNugget would account for about 10 percent of McDonald's total sales, an amazing transformation for a company made famous by beef.[36] As Eric Schlosser wrote in *Fast Food Nation*, "The Chicken McNugget turned a bird that once had to be carved at a table into something that could easily be eaten behind the wheel of a car. It turned a bulk agricultural commodity into a manufactured, value-added product."[37] It would also help transform Tyson Foods into the largest poultry producer in the world.

The size of the McDonald's venture was only part of the story. With the McNugget, the most important restaurant in the world—one known worldwide for its burgers—was launching a massive advertising campaign to promote a product it had little previous association with. The leap into chicken made perfect sense. By 1983, the question was not *if* chicken would surpass beef as America's favorite meat, but *when*. McDonald's, which was clearly threatened by declining levels of red meat consumption, jumped on (and then turbo-charged) the chicken bandwagon. In the process, it helped turn a healthy source of protein into one of the fattiest items on its fat-filled menu. McDonald's initially claimed its McNuggets were made from "whole breasts and thighs," but it turned out that they included processed chicken skin and were fried in saturated beef fat.[38] When McDonald's introduced its "all white meat" chicken nuggets in 2003, many consumers couldn't help but ask: What were these things made of before?

In the end, it didn't matter: McNuggets were not red meat, which by the 1980s had been blamed for just about every nutritional sin. Americans, particularly American children, gobbled up McNuggets by the millions, and the chicken was forever transformed. With increasing frequency, the nugget, or some close cousin, became the standard form of presentation and preparation.

By the mid-1980s, sales of chicken were the fastest growing in the fast-food industry, and fast food was the most rapidly growing part of the restaurant industry. More Americans were eating out than ever before, and they were eating in fast-food chains. Kentucky Fried Chicken was the undisputed chicken king, with Colonel Sanders the second-most-recognized public figure in the world.[39] But other chicken-centered restaurants such as Church's, Popeye's, Chick-fil-A, and Bojangles dotted America's landscape. Initially, the chicken chains viewed the nugget as beneath them and were a bit slow in introducing their own versions. In the end, however, the product's success made converts out of everyone. The nugget was not simply a new form of chicken; it was a new market. Working women and especially children loved the little chunks.[40] Poultry producers, in turn, not only produced them for fast-food restaurants but also marketed them to supermarket shoppers.[41]

More importantly, the giants of the fast-food industry, McDonald's and Burger King, saw chicken as their future. The introduction of a chicken sandwich by McDonald's is thought to have increased the nationwide demand for chicken breasts by some 6 percent in 1989.[42] The Wendy's catchphrase "Where's the beef?" from the early 1980s may have kept us laughing, but when we went to this chain to eat, more and more of us were ordering its wildly successful line of chicken sandwiches.[43]

Chicken's supremacy went well beyond fast food. Trends set in food service, whether restaurants, cafeterias, or banquet halls, quickly filter down to the retail world of supermarkets, delis, and the dinner table. About 40 percent of chicken is sold in restaurants, 40 percent is moved by retailers, and another 20 percent is now shipped abroad (a fairly recent development).[44]

Prepared chicken dinners, from companies like Banquet Foods, Swanson's, and Weaver, took off in the 1980s, as did easy-to-prepare lines of chicken nuggets, sticks, and soups. Mimicking the taste and convenience of restaurant food, supermarkets and their suppliers struggled to keep their increasingly mobile consumers eating at home. Processed convenience foods quickly became a familiar sight in our refrigerators and freezers.

Chicken now occupies the center of the plate regardless of whether that plate is served at home or in an airplane, restaurant, retirement home, or school cafeteria. A cheap source of protein with a healthy image, chicken has the additional advantage of being exceptionally versatile. Chicken is processed and sold in more varieties than any other meat, from bone-in chicken for frying, baking, roasting, grilling, and rotisserie prep to boneless nuggets, tenders, fillets, patties, chunks, and meatballs.[45] Chicken is routinely used in salads, wraps, sandwiches, and stir-fries. And "ethnic" foods, such as Mexican, Chinese, Thai, and Indian, have all recognized chicken's remarkable versatility and consumer appeal.

Chicken is also remarkable for what it can absorb and retain. According to an industry periodical: "Chicken's versatility makes it a perfect carrier for today's most popular flavors. Seasoning options range from the simple addition of a little salt and pepper to more complex combinations of liquid and encapsulated flavors and seasonings, flavor enhancers, spices, and oleoresins. Seasoning systems can be applied to poultry topically or internally, depending on the application and desired flavor impact. Internal application of marinades uses vacuum tumblers or injection systems. Processors typically inject seasonings in bone-in poultry products and marinate those without the bone. The object is to get the flavor system inside the meat or as close to it as possible for maximum flavor retention. Topical flavor systems include rubs, glazes, and sauces."[46]

Is this what Herbert Hoover had in mind?

The irony is that branding and processing have undermined the reasons why so many folks turned to chicken in the first place: its low cost and healthfulness. These are still the reasons that many of us list for buying chicken. But in the end our "choices" tell another story. If Americans were "switching" to chicken for purely health reasons, we should have seen a growth in the consumption of whole birds and a decline in hamburger meat. This is not the case. Rather, statistics show that since the mid-1970s we have moved away from red meat by substituting processed chicken for higher-quality table cuts. This shift has not been as healthy or as cost-effective as we like to think, and it has been fueled by the proliferation of processed chicken products during the past two decades (combined with our belief that chicken can do no wrong).[47] Only whole chickens or store-brand thighs beat hamburger in terms of cost. Once "value" is added to the original product, chicken becomes more expensive than ordinary hamburger.[48]

Perhaps more importantly, processed chicken is often less healthy than red meat. This is especially true in the case of fast food, a branch of the industry that has in large part defined the way we eat chicken. Six chicken nuggets, not exactly a filling meal, contain the same amount of fat (twenty-one grams) as a fat-filled double cheeseburger with condiments and vegetables. Order a chicken fillet sandwich and you get about one-third *more* fat than that same burger.[49] In some cases, such as the Burger King Chicken Sandwich, it is even worse. With forty-three grams of fat and over seven hundred calories, it is less healthy than almost any burger.[50] To be sure, there are alternatives. Grilled chicken sandwiches are relatively healthy, but fast-food varieties tend to have little taste and are served in oversized portions and with unhealthy sides. Burger King's Grilled Chicken Whopper, for example, which is probably the tastiest of the bunch, has almost six hundred calories and fourteen grams of fat.[51] Add fries and a shake and look out.

Worse yet, the healthier alternatives seemed to get less healthy during the 1990s. McDonald's grilled chicken sandwich put on 140 extra calories from a thick layer of mayo, while its Grilled Chicken Deluxe started out with three grams of fat but now has a whopping twenty grams. But even this hefty sandwich is surpassed by Burger King's BK Broiler, with its astonishing twenty-nine grams of fat.[52]

Unfortunately, non-fast-food chain restaurants do not serve much healthier fare, especially to children. The Center for Science in the Public Interest found that fried chicken fingers or nuggets were central to each of the kids' menus it surveyed recently. Chili's innocent sounding "Little Chicken Crispers," for example, delivers (in three small pieces) 360 calories and eight grams of the worst kind of fats (saturated and trans).[53]

Endorsed by a matronly celebrity, a famous athlete, or a movie star, and then placed into flashy packaging with a Tyson, Burger King, or McDonald's logo, the chicken we eat today is frequently more expensive and less healthy than the unhealthy product we intend it to replace. In this way, a cheap, healthy food has become a not-so-affordable way of getting fat.

II

An American Industry

The rise of chicken as America's favorite meat has been fast and dramatic. At the millennium, the average American was eating over a hundred times more chicken than a person was eating on the eve of the Great Depression.[1] How did this happen? Why did the broiler industry take off in the 1940s and 1950s, and then continue to expand during the latter part of the twentieth century? What factors led the industry to settle in the South? And who, in addition to health-conscious, convenience-oriented consumers, drove this expansion?[2]

The Early Days

Until the mid-1920s, the raising of chickens for meat was hardly a central business for most American farmers. Women on American farms often raised a small flock for home consumption and a little spending money. Methods for handling the birds were primitive. Chicken houses that cost $200,000, automated food and water dispensers, and carefully calculated feed

Feeding chickens in the 1930s
(Photo by Wes McManigal, Grant Heilman Photography)

mixtures were fantasies of the future. Chickens still roamed the farmyard and were fed excess grain and table scraps. That was the beauty of it. Why would anyone seek to change a system in which feed and labor were essentially free?[3]

In regions with good soil, flat land, and favorable weather, such as the Midwest, it made little sense to devote serious resources to chickens when farmers raised them for next to nothing—instead, such farms specialized in more profitable enterprises such as corn, wheat, hogs, and cattle. In less-hospitable farming regions, however, farmers were willing to try just about anything. Poverty, then, is the mother of the modern broiler industry.

It was in the Delmarva Peninsula (the eastern shore of

Delaware, Maryland, and Virginia) where the broiler industry
first emerged in the 1920s. According to legend, Mrs. Wilmer
Steele of Delaware was the first to raise chickens solely for market. In 1923 she ordered and raised five hundred chicks. Sales
were strong, so the next year she tried a thousand. As the business became even more profitable, her husband left the Coast
Guard and started building chicken houses. At this point, word
spread to other enterprising farmers. From a modest beginning of only 50,000 birds raised in 1925, the state of Delaware
produced annually one million birds by 1926, two million by
1928, and three million by 1929. By 1934, Delmarva farmers
were raising around seven million birds a year, and the region
was the undisputed poultry capital of the world.[4]

Two factors explain Delmarva's remarkable rise. First,
farmers in the region had been devoted to the highly risky business of farming table vegetables, which had to be sold quickly
and locally. Decent years could be followed by crop failures
and economic disaster. Delmarva farmers turned to chicken in
the hope that it would provide a steady, if not spectacular, source
of income. Raising chickens would allow folks to get through
the lean times.

Second, Delmarva was located near all of the major East
Coast markets, particularly Philadelphia and New York, but
also Boston, Baltimore, and Washington, D.C. At the time,
chickens had to arrive at the market alive. Consumers bought
recently slaughtered birds from their local butcher. Because of
the difficulties involved in transporting live birds, growing regions had to be located near consumers; thus Delmarva was
ideally situated.

Although the industry lacked coordination and planning
during these early years, it expanded rapidly. Individuals became involved in the business almost by accident. Homer Pep-

per, for example, of Selbyville, Delaware, bought a Model T in 1921 to haul ice cream. One day he had to make a trip to Philadelphia and decided to take along eight coops of chickens to sell in the market—thereby starting the business of transporting chickens: "It was right there that the business of transporting chicken got its start. He made enough off that first load to warrant further loads. Most of the chicken hauling was done during the winter months. During the summer, Mr. Pepper hauled produce to the markets. This combination load proved profitable enough that by 1924 he operated two trucks and a trailer, with 60-coop capacity each. Road conditions were so bad that one trip to Gumboro a week, for example, to pick up chickens, was about all that could be managed."[5]

By the mid-1930s, Delmarva farmers were raising so much chicken that buyers from New York and Philadelphia came to them. The market was still competitive, and growers could sell their birds to the highest bidder. Not surprisingly, then, within a few years some of the first processing plants began to emerge in Delmarva. Birds were "New York–dressed": slaughtered and defeathered but left with the entrails and feet intact. Although it was a long way from nuggets and patties, this basic processing would eventually transform the industry. Birds no longer had to be transported alive; they could be processed near growing areas, then packed on ice and trucked to market. This allowed growers in Delmarva to ship their chickens to more distant markets, but it also reduced their competitive advantage. Indeed, the advent of processing opened up the possibility of chicken farming to rural folk throughout much of the United States.

Prior to 1940, however, Delmarva was in a league of its own. It remained the only region in the country with anything resembling a broiler industry—where, in some uncoordinated way, hatcheries, farmers, feed dealers, and truckers turned chicks into broilers and sold them in distant markets. Outside of Delmarva, the poultry market was purely local. Live chickens left farms on trucks and were sold in nearby markets.[6] Processing plants, which would eventually expand the geography of the industry, were not widespread until after the war. Most regions simply did not have enough chickens or capital to warrant the creation of processing facilities. Like virtually every other region in the South that would eventually experience a chicken boom, the Ozark Mountains had never enjoyed a golden age of farming.[7] As one old-timer put it: "It was never easy here for the farmer. Don't let anyone tell you that. My family [in the 1930s and 1940s] raised cattle, grew corn, had apples, logged trees. . . . We did anything we could to make money. Everyone was like that. . . . The land just isn't that good. Those who did the best were always the ones who found a way to get out of farming . . . own a store, drive a truck, or whatever. It's always been that way. When people started raising chickens no one thought much of it. It was just another way to make money. And it was risky, just like everything else. It's never been easy here . . . even after chickens."[8]

The common feature of future poultry-producing regions in the South was poverty, enduring poverty. Low-yielding and hilly land, few resources, and limited access to credit kept people scrabbling to simply exist. As a longtime resident of western North Carolina put it: "Working people always have to struggle to survive. This is true everywhere. But this place was a bit different. Here, in the 1930s and 1940s, *everyone* was poor and no one seemed to ever get out. To get out you need something. People here had nothing. . . . The lucky ones fought in the war. The rest were stuck here—*for generations.* Some ran moonshine. Others had a little farm. Most just scrambled."[9]

In the 1930s, few people in northern Georgia, western North Carolina, or northwest Arkansas thought that there was much future in chicken. Not even "pioneers" like John Tyson or Fred Lovette had a clue as to how big this little bird would become. At the time, there were three interrelated, and apparently intractable, problems that seemed destined not only to prevent chicken from taking a hold in the South, but also to keep it from developing into a full-scale industry.

MARKETS

Local markets in the South were too small to stimulate and sustain the industry, and it was difficult to transport live birds to larger population centers in Chicago, Miami, the Northeast, and the West Coast. Assuming that chickens could be produced and consumed in large quantities (two big ifs in 1930), how would live birds be kept hydrated, fed, and breathing as they moved from farm to market? The process was not only hampered by poor southern roads and primitive methods of transportation, but also severely complicated by a lack of coordination among the various links in the chain (growers, feed dealers, truckers, and butchers).

PRODUCTION

Could farmers raise large numbers of chickens both cheaply and profitably? Chicken had less status than beef, but when purchased at the market it cost as much as lobster. If Americans were to start buying chicken, the price had to be drastically reduced. Initial drops in production costs could be expected once farmers saw chicken as a serious business and attempted to raise the birds more efficiently. The extent of initial improvements was limited, however. Farmers, especially those who were poor enough to be attracted to chicken in the first place, had few resources. They could not be expected to fund, let alone conduct, the scientific research needed to improve chicken breeding, housing, feed, health care, and so on. Would "big science" step up to the plate?

CONSUMPTION

If farmers raised chickens and then if middlemen got them to market, would consumers buy them? From a modern-day perspective that includes Buffalo Wings and Chicken McNuggets, the answer seems obvious. Prior to World War II, however, there was no guarantee. Delmarva's emergence in the 1930s was impressive. The region was producing about two-thirds of the country's broilers. But at the time Americans were still eating only a few pounds of chicken a year. If chicken was to go from "Sunday dinner" to an everyday food, its cost had to be reduced and its image improved. Although expensive when purchased in the market, chicken lacked cachet. In the country, chickens were the lowliest of farm animals. In cities and towns, they were a backyard nuisance. Americans would eat chicken, but they were not willing to pay for lobster when they were getting, well, chicken. Early initiatives suggested that markets in the East Coast, Midwest, and West Coast were far from saturated, but it was unclear how much chicken Americans would eat.

It was in this uncertain context that John Tyson moved from Missouri to Springdale, Arkansas, in 1931 with his wife, his one-year-old son, Don, a "nickel in his pocket," and a truck that allowed him to run a small business hauling fruit, hay, eggs, and, eventually, chickens: "And that is all I had. I wanted a cup of

coffee, and I had the money to pay for it. I'll never forget toss-
ing that nickel down on the counter. I tried to make it look as
though I had a bankroll in my pocket to back it up. Then I
went out and started looking for a load to haul somewhere."[10]

By 1935, Tyson and a number of other entrepreneurs were
regularly transporting chickens from northwest Arkansas to
Kansas City and St. Louis. At the time, these were the only major
markets that could be reached from Arkansas (if the birds were
to arrive alive). Tyson was continually improving his fledgling
enterprise, however, and soon developed a contraption that
allowed the birds to eat and drink while they traveled to market. This invention brought more profitable markets into reach,
and in 1936 Tyson took his first load of birds to Chicago. During
the next year, he made similar trips to Cincinnati, Detroit,
Cleveland, Memphis, and Houston.[11] With a greater range of
markets now accessible, Tyson focused on the most basic of
problems: How could he get more birds?

John Tyson was not a farmer; he was a middleman, and
hauling birds gave him an intimate understanding of all segments of the emerging industry. When he lacked chicks to deliver to his growers, Tyson bought a small hatchery. When he
had problems accessing feed, he became a feed dealer for Ralston Purina and eventually built his own commercial mill. In
this sense, the process of vertical integration, whereby previously independent facets of the emerging industry were brought
under the control of a single entity, initially occurred as a response to problems encountered along the chain of production.

In fact, most chicken industry pioneers owe their early
success to roles in transportation. As we will see in Chapter 3,
Charles Odell Lovette, the patriarch of the family that was to
build North Carolina–based Holly Farms into one of the largest
poultry companies in the world, began in the mid-1920s by

gathering country produce in his Model T and hauling it to city markets. Similarly, at around the same time, Jesse Jewell, one of the industry's first integrators, began to drive around the hill country of northern Georgia promoting to local farmers the idea of raising broilers. Most farmers in the region were struggling to raise cotton on poor land, an enterprise that during the Depression did little except increase their debt. To make matters worse, a 1936 tornado wiped out much of Gainesville, Georgia, the future poultry capital of the world. It was in this context that Jewell tried to revive his mother's failing feed business. To address the problems of farmers having little livestock or money with which to purchase feed, Jewell placed day-old chicks with farmers and extended them feed on credit. When the chicks became broilers, he picked them up, settled with the farmers, and hauled the birds to Miami. Farmers in northern Georgia embraced the system in part because they had been using it for their other crops. A local store would furnish feed and fertilizer, along with a little store credit. When the crop came in, it was sold back to the store and the accounts were settled.[12]

Jewell's operation, as well as northern Georgia poultry, took off from that point. He quickly invested in a fleet of trucks, which allowed him to expand his operations, bringing feed, chicks, and other supplies to his growers while delivering mature birds to expanding markets. But he didn't stop there: "In 1939 Jewell built his own small processing plant in Gainesville to clean, eviscerate, and freeze his broilers, readying them for marketing by a wholesaler. In 1940, to capture economies of scale in chick hatching, he built a hatchery that turned out 12,000 . . . chicks per week. Jewell was constructing an integrated broiler system, and the northern Georgia hill-country bounded by Gainesville, Cumming, and Canton was coming to be known as the 'Chicken Triangle.'"[13]

By World War II, then, Jewell had one of the first vertically integrated poultry companies in the country. He controlled the grow-out phase through contracts with farmers, and was establishing control or ownership over baby chicks, processing, transportation, and marketing. A decade later, in 1954, the company could boast that "with the completion of their new feed mill, J. D. Jewell will have forged the last link, giving them complete step-by-step quality control over the production of frozen poultry." Jewell now controlled laying flocks, incubation, grow-out, warehousing and distribution, sales and advertising, by-products, and processing. Along the way, he was one of the earliest to abandon New York–dressed in favor of frozen chicken. By eliminating spoilage, the frozen chicken allowed Jewell to stabilize both production and supply and to gain access to markets throughout the East Coast. Indeed, by the mid-1950s, Jewell had built the largest integrated poultry company in the world.[14] According to a local newspaper, the "fuzzy down of the baby chick has all but ousted the fleecy lock of the cotton boll from its pedestal as chief money crop of Hall County."[15] Cotton was no longer king. Welcome to the New South.

As with the industry's other pioneers, Jewell's rise to the top of the poultry world was not without bumps. In one case, after workers cast their ballots for union representation at one of Jewell's plants in the early 1950s, they were "viciously attacked and beaten by a mob wielding blackjacks and rubber hoses. Participants in the mob included top level company supervisors and company employees, acting by permission and with the approval of company supervisor representatives." Jewell, who paid his workers seventy-five cents an hour "for all types of work, regardless of the length of experience," was out-spokenly antiunion and worked to keep Gainesville union-free. He also was not above disciplining "his" workers. Unfortunately, Jewell's methods for handling farmers and workers would become industry standards.[16]

By the 1940s, then, many features of the current industry were beginning to take shape. Signs of integration were on the horizon, with pioneers Jesse Jewell, John Tyson, Fred Lovette, and others beginning to extend their operations into all segments of the industry. But even as late as 1950 there were thousands of specialized mom-and-pop chicken operations existing alongside large feed companies and budding integrators. In short, there was still competition all along the chicken chain. Industrywide integration would be a 1950s phenomenon.

It was also clear by the 1940s that there was a rural farming population willing to raise chickens. To be sure, farming families were not getting rich off of chicken, and more than a few wondered whether their small businesses would survive from one year to the next. Prices for feed and broilers fluctuated tremendously. As one early farmer remembered: "It was not unheard of for a grower to take feed on credit, raise a batch of chicks, and then sell the broilers for less than what he or she owed on the feed. You'd be better off working for free! You could raise a whole batch of chickens and then owe money for your trouble. What a deal!"[17]

Although the emerging system of contracting would eventually protect farmers from the market, raising the birds was an inherently risky enterprise. Once farmers began to keep large numbers of chickens in small places, they required modern methods of confinement, waste management, disease control, breeding, and nutrition. The industry lacked uniform stan-

dards of control and ways of disseminating information to ensure that all those involved in the chicken industry were receiving and delivering high-quality products (from chicks and
feed to medicine and transportation). In short, although the
science and state intervention necessary to safely support the
growing industry were beginning to take shape, there was a
long way to go. Further, investing resources in these supports
still offered an uncertain payback. It was unclear both whether
chicken farmers would survive long-term and whether Americans would consume enough chicken to make the business
profitable.

World War II and the Industrial Chicken

It is hard to overemphasize the profoundness of the changes
within the poultry industry during the 1940s and 1950s. By the
end of the 1950s, Americans were eating an astounding amount
of chicken compared to the prewar period, large integrators
were firmly in control, the locus of the industry had migrated
South, and the biological productivity of the bird had been
revolutionized. All of these changes are directly connected to
World War II.[18]

World War II put chicken on more American dinner plates
than ever before. Unlike beef, chicken was not rationed during
the war, and the federal government set a price that was well
above the cost of production. Even the ceiling price of thirty
cents a pound made chicken profitable, and the black market
price of fifty cents made chicken well worth farmers' while. If
that wasn't enough support, the federal government intervened
in at least two other important ways. First, through its "Food
for Freedom" program, the government encouraged consumers
to eat eggs and chicken in order to leave more "desirable"

Japanese-American in relocation center during World War II
(Photo by Ansel Adams, Library of Congress)

sources of protein such as beef and pork for the troops.[19] Raising and eating chicken were now patriotic duties and a matter of national security. Second, in 1942, the War Food Administration commandeered all broilers coming out of the Delmarva Peninsula for federal food programs. At the time, Delmarva was producing around ninety million broilers a year. No other area came even close to this number; Georgia and Arkansas, the future kings of the industry, each hovered around ten million birds grown annually.[20] This wartime policy effectively meant that the premier poultry-producing region in the country, a region that produced over half of the country's commercial broilers, was suddenly removed from the market. The mandate not only increased demand; it also reshaped the geography of the industry by opening the country's major markets to producers outside of Delmarva. This was all the opportunity that southerners like Jewell, Tyson, and Lovette needed. By the time the war was over, the geographic transformation of the industry was irreversible. To this day, Delmarva remains a major producer, but it never recovered the markets it lost during the war or regained its place at the top of the industry. By 1950, the South was the most dynamic broiler-producing area in the United States.[21]

The rationing of beef, the regulation of prices, and the dramatic increase in demand during World War II all helped make chicken an everyday food for American consumers. American broiler production almost tripled during the war, increasing from 413 million pounds to 1.1 billion pounds between 1940 and 1945.[22] And that was just the beginning. What the country wanted was meat that was better, cheaper, and more plentiful. The two interrelated forces of science and integration, along with growing demand and cheap labor, delivered just that during the late 1940s and 1950s. The industrial bird was born (or rather, created). William Boyd sums up the process: "Beginning in the interwar years and accelerating rapidly after the Second World War, advances in nutrition, health, and genetics translated into massive increases in the biological productivity of broilers. Such gains facilitated and were in turn reinforced by the subsequent integration of the industry. . . . At the same time, the incorporation of hatcheries, feed mills, contract grow-out operations, and processing plants within a single firm provided an institutional vehicle for further rationalizing the production system in order to capture productivity gains. As the industry grew in size and sophistication, moreover, there was a clear shift in the locus of research and innovation from the public to the private sphere. By the early 1960s, integrated firms, primary breeders, and animal health companies had become the drivers of innovation in the industry, transforming the lowly chicken into one of the more thoroughly industrialized commodities in American agriculture."[23]

Simply put, "the barnyard chicken was made over into a highly efficient machine for converting feed grains into cheap animal-flesh protein." And how. "Between 1935 and 1995 the average market weight of commercial broilers increased by roughly 65 percent, while the time required to reach market weight declined by more than 60 percent and the amount of feed required to produce a pound of broiler meat declined by 57 percent. In short, a commercial broiler from the 1990s grew to almost twice the weight in less than half the time and on less than half the feed than a broiler from the 1930s."[24]

The quest to create the genetically "perfect" bird continues today.[25] Nevertheless, the foundation for better breeding was laid in the period immediately following World War II and was strengthened by public investments from land grant universities and the federal government in nutrition, disease control, and technologies for housing chickens. A series of factors, most notably better feed (now with vitamins and antibiotics) and housing for the birds, led to lower mortality rates and higher feed conversion ratios while reducing labor needs. In 1940, for example, 250 person-hours were required to raise a thousand birds; by 1955 that time had been reduced to 48 hours.[26]

Advancements in one segment of the industry, however, could not be fully realized unless improvements occurred all along the chain. Productivity increases expected from a better breed or a higher-quality feed could not be fully realized if methods for confining the birds, controlling disease, and disposing of waste were still primitive. The greater need for coordination facilitated the push toward integration, whereby a single firm came to control, standardize, and monitor all of the inputs and outputs necessary for breeding, hatching, feeding, housing, growing, and processing large quantities of birds.

Industry integration during the 1950s and 1960s was defined by the growing power of large national feed companies such as Pillsbury and Ralston Purina. In the 1940s, contracts were fairly informal, with growers paying for chicks, feed, supplies, credit, and other inputs after the sale of the birds. Feed mills, often working on credit from banks, extended credit (in the form of feed) to local feed dealers, who in turn extended credit to growers. The problem, however, was that broiler prices fluctuated significantly. When prices rose, growers produced more broilers, thereby increasing supply and lowering prices. Farmers who were then unable to make their payments caused disruptions along the entire chain. To secure their investment, large feed mills increasingly began to bypass the feed dealer and sign contracts directly with farmers.[27]

By the mid-1960s, integration was complete. Integrated operations, run primarily by large national feed companies like Ralston Purina and Pillsbury, accounted for 90 percent of broiler production. These firms tended to own hatcheries, feed mills, and processing plants, and to contract farmers for the grow-out phase. Only foundational breeding, which by the 1960s was controlled by a handful of companies, and consumer sales, which was left in the hands of supermarket and restaurant owners, remained outside the direct control of the large integrators.[28]

The emergence of extremely powerful, vertically integrated firms, along with the spread of contracting, meant different things for different people. For integrators, the system greatly improved efficiency and stability. By controlling virtually everything that goes into the raising and processing of chickens, they were able to ensure that a high-quality broiler was produced cheaply and in extremely large quantities. Most integrators felt then, and still feel today, that the system was in the best interests of growers as well. To be sure, integrators recognize that growers make considerable investments, furnishing around half of the industry's working capital (in the form of poultry houses, equipment, and so on). But they insist that the system of contracts guarantees growers some profit while effectively insulating them from market fluctuations. In the late 1950s, during this period of intense vertical integration and industry consolidation, Jesse Jewell put it this way to the U.S. Congress: "Our operation is all contracts, practically 90, 95, maybe 100 per cent. These farmers down there won't grow chickens unless they know they are not going to lose anything, so our company, in order to control production—and we like it that way—in order to have the regular supply of chickens coming in, we let them feed. We furnish the baby chicks and the feed. [The growers] furnish the house, the equipment, and the labor, heat, and we pay them according to the number of pounds of chickens they get out of the number of pounds of feed they use, and it is a flexible contract and as things get a little tough we can sort of twist up the contract and as they get better we have to loosen up the contract in order to hold our growers."[29]

It is interesting to note that these statements were made during a period when integrators such as Jewell had to "hold" their growers. Growers then operated in regions that were serviced by a number of integrators. If Jewell did not treat his growers properly, they would switch integrators. Today, very few growers have that luxury. On a local level, integrators generally enjoy a complete monopoly, having effectively agreed to stay out of each other's way—a situation perhaps anticipated by one grower expressing his concerns to Congress in 1961: "I wish to impress upon the committee the fact that severe pressure and controls are being imposed on the grower by the feed companies and the processors. The role of the family-size poultry producer has been reduced to 'a cheap hired hand with a large investment.' The feed companies have a monopoly of contracts over the entire industry. . . . They make just enough variation in the contracts to avoid being prosecuted for price-fixing. If the grower disagrees or objects to their demands, they refuse to deal with him, leaving him with a large debt for buildings and equipment . . . We are a depressed people, and our only hope is through some type of legislation."[30]

Contemporary growers echo such concerns, often in quite blunt terms:

> I ask you: Who is worse off, the plant worker or the poultry grower? We are. The poultry grower. No doubt. Sure, plant workers know a thing or two about hard work and low pay. I worked in a plant. It's no picnic. What makes [the growers' situation] worse is that we are trapped. You can quit a job. You can't quit raising chickens. Imagine if you went to get a job in a plant and the supervisor said, "Sure, we'll give you a job cleaning up chicken guts, but first take out a loan for $200,000"? Who would do that? Well, that is what they say to growers. And we do it! We take on a big debt in order to finance the houses and all the equipment. Once you have

the debt you are trapped. The only way you can make your payments is by raising more chicks. Tyson knows this. So you can't say boo. If you complain too loud, Tyson will just stop bringing you chickens, or give you lousy chicks, or screw with your feed. Then you have houses that aren't good for anything and a ton of debt. Sure, when you start, you think: I'll pay off this debt in ten or fifteen years and after that it will be all profit. But there is a catch. Every few years the company requires "improvements"—improvements to the house, a new this or a new that. These "improvements" are really experiments. Most of the time the company has no clue if they will make any difference, or if the difference warrants the investment. And they certainly don't care if it improves *our* bottom line. It's all about profit for the company. Not only do we have to do these experiments to our businesses, we have to pay for them! So we take on more debt. All for the company. Have you ever heard of anything so crazy?[31]

To suggest that growers are all of one mind would be misleading.[32] Their opinions about integrators, contracting, and the business of raising chickens vary tremendously. One grower, for example, told me:

Some say growing chickens is like being a slave or sharecropper. I'm sorry, I just don't see it that way. Sure, in the end, the [integrator] has more control. Almost total control really. No one can deny this.

But the grower still has independence, not from the
integrator. There is always somebody above you.
But the way we live is independent. Chickens let me
keep the farm, be with my family. And there is no
boss breathing down my neck. I get up when I want,
go where I want, work when I want. And I make a
decent profit. It's hard if you try to live just off
chickens, but if you have [other sources of income]
then chickens help out. That's what the company
says, and I agree. I don't like everything about the
company, but this is a good deal, the best you can
get in this area. The money is consistent. And I like
raising chickens. Maybe that sounds strange. But
I'll tell anyone. I *like* raising chickens.[33]

It is tempting, when faced with such diverging opinions,
to place growers into two opposing camps: pro-integrator suck-
ups versus anti-integrator troublemakers. In this particular case,
the happier grower was in the fortunate position of having a
large operation (six poultry houses), amazingly little debt, and
considerable off-farm income. A rare combination, indeed. But
growers' understandings of the industry are not mechanically
tied to their economic situation any more than their opinions
can be simplistically labeled pro- or anti-integrator. Growers
have differing resources, histories, hopes, and perceptions.[34]

Once vertical integration and contracting became the industry norm, however, everyone—regardless of their views—
was subject to that system. Subsequent changes within the
industry, including further consolidation, mergers or acquisitions, and corporate shake-ups, all occurred within this basic
framework. In the late 1960s and early 1970s, continued insta-

bility within the industry led many of the feed companies to get out of poultry. This allowed regional chicken-centered integrators like Tyson, Perdue, and Holly Farms to emerge as the industry leaders during the 1970s and 1980s.[35] These chicken-centered companies would not only take over the industry; they would transform it by changing how chicken is presented to the American consumer.

III
Anatomy of a Merger

It was never easy for the grower. In the early days, there was no security. None. Sometimes you would get chicks, other times you couldn't find chicks anywhere. It was really competitive. Not just one integrator like today. [And] it was chaotic. Things were always changing. Someone would deliver you chicks or feed one day, and be gone the next. And the prices fluctuated so much that sometimes you really were working for nothing. You were independent, but there was no security.

Guess what? It's hard to believe, but the situation is actually worse today. We traded the chaos of the past for the total control of the integrator. That's what I say. We went from one extreme to the other. Today there is very little uncertainty. I get my chicks, feed, and everything else like clockwork. But in exchange I am a slave to Tyson. [Growers] invest $150,000 a house, watch the chickens twenty-four hours a day, and do everything

and anything Tyson says. Tyson is the only game in town.
It's the only integrator around here so there is no choice, no
competition. If Tyson wants improvements, you make them,
you pay for them, and you smile real nice. And in exchange, if
you keep your mouth shut and work hard, you keep getting
chicks delivered so you can pay the mortgage. What a system!
Have you ever heard of anything so crazy? We are the only folks
in the world who actually pay—go into debt—to be slaves.
That's the history of the poultry industry.

Those few growers who have been around long enough to remember the brief moment before the industrywide spread of contracts are hardly nostalgic. There have been no golden years for poultry growers. Contracts and integrators are seen as inevitable, even beneficial, developments. Nor do growers hold anything against local big shots—as long as they did not forget their roots as they transformed mom-and-pop operations into corporate giants.

What growers are critical of are the contract terms that have developed over the years and the changing context in which contracts are "negotiated." For most growers, the greatest problem with the industry is the shift from an economic playing field characterized by a multiplicity of "local" firms to one controlled by a handful of extremely large, impersonal, and fully integrated corporations (whose ties to local communities are contrived by public relations departments). It's not that the "good ol' boys" were ever that good, but they were "ours." They were accessible and could even be held somewhat accountable. More important, they operated within a context of competitive capitalism: "Early on, if a company treated growers like shit the grower just went somewhere else. There were choices. Today there are no choices."[1]

The Rise of Holly Farms

In 1924, Charles Odell Lovette built a cover on his Model T and began buying chickens, eggs, and produce from local farmers living around Wilkesboro, a small town in the foothills of the Blue Ridge Mountains in western North Carolina. At the time, Wilkes County was the heart of moonshine country, a status that reflected the enduring poverty and lack of economic options in many southern hill regions prior to World War II. The Lovette family business grew slowly during the Depression, and by the mid-1930s the oldest son, Fred, began running one of the truck routes. Fred Lovette would become to Holly Farms what Don Tyson would become to Tyson Foods. Their fathers, Charles Lovette and John Tyson, were the founders and patriarchs. But it was Fred and Don who were the innovators and risk takers, building empires that would make their respective companies synonymous with chicken.[2]

When Fred graduated from high school in the early 1940s, he immediately borrowed $3,100 from his father, opened up an office in North Wilkesboro, and started the Lovette Poultry Company. Each of the Lovette men, with the help of their wives and children, focused on a separate facet of the emerging industry. Charles concentrated on eggs, a younger brother ran the live end of operations, and Fred was in charge. Lovette Poultry grew slowly. Labor shortages during the war, a lack of capital, and poor roads all kept the expansion of the Wilkes County poultry industry in check. The Lovettes were successful by local standards, trucking live chickens to markets in Charlotte,

Greensboro, New York, and Chicago, but they remained a small marketer of live birds through the 1940s.[3]

In 1947, the prospects for the Lovettes expanded dramatically when three local GIs, Harry P. Hettiger, Forrest E. Jones, and Vernon Deal, decided that the freezers arriving in American homes should be filled with frozen dressed chickens ready for the frying pan. Together, they incorporated the Wilkes Mountain Poultry Products Company and purchased the feed warehouse of Holly Mountain Farms in downtown Wilkesboro. The feed warehouse had belonged to Harry's brother, E. P. Hettiger, the general manager of Tuxedo Hatchery Inc., which included both a feed department and large breeding farms. Within a year, the former warehouse was, according to the *Winston-Salem Journal and Sentinel,* one of the most modern poultry-processing plants in the world. The plant, where "efficiency is a corollary of religion" and "deft-fingered girls" worked with amazing speed, employed fifty people and supplied about a million frozen fryers a year to customers in Kentucky, the Carolinas, and the Virginias.[4] At the time, few poultry-producing regions outside Delmarva had the right combination and density of markets, growers, and capital to warrant such a processing plant. With the arrival of the plant in Wilkes County, the area's poultry industry was suddenly on the map. According to a local newspaper: "Poultry is said to have afforded the Wilkes people an income of $7,000,000 in one recent year. It's not hard to believe. It may be believed, too, that this processing plant is ushering in a new era in the county. In this era the poultrymen may be expected to stand together and work for the good of their industry. The owners of hens will supply the hatcheries; the hatcheries will supply the producers of fryers with baby chicks and feed; the producers will supply the processors; the processors will supply the grocer; and the ultimate consumer will be better served than ever."[5] Wilkes County had indeed entered a new era. A modern poultry industry, with all the components needed to breed, hatch, feed, raise, process, and market broilers, had been established.

The continued success of the industry was, however, hardly assured. The processing plant, now under the control of E. P. Hettiger and renamed Holly Farms, struggled. When prices were high, farmers raised more birds, flooded the market, and drove prices down. Growers would then stop producing, causing the supply to dry up. It was not uncommon for a farmer to start a flock and be forced to sell after prices had dropped, thereby preventing him from starting another batch. This was a real problem for Hettiger. He needed a steady supply of cheap chicken to keep his plant profitable.[6]

Fluctuating supply was also a problem for Fred Lovette because if he could not meet the demands of his buyers, they would turn to other sources. As a result, in 1948, Lovette, now teamed up with "Izzy" Kendrick, a poultry man from Virginia, did what no one in Wilkes County had done before: formed a feed company and began contracting with growers to ensure a steady supply. Lovette Feed furnished chicks and feed and guaranteed the farmer would take no loss.[7]

Resolving the problem of supply through contracting proved key in Wilkes County and the industry as a whole (see Chapter 2). At this early stage, the system was relatively informal and appeared to have no downside. It generated a steady supply of chickens, which helped everyone, and insulated farmers from risks associated with the market (which kept them in business). As contracts became more rigid, and fewer and fewer corporations ruled the roost, the system would become increasingly disadvantageous for the grower. In Wilkes County in the mid-1950s, however, the long-term impact of contracting was far from clear. People were just trying to survive in the short-term.

With contracting, Lovette could now guarantee a steady flow of broilers, and he decided to buy the processing plant from Hettiger. By January of 1955, then, Lovette was the sole owner of Holly Farms, a company that now employed three hundred people and had a production capacity of 175,000 birds a week. Later that year, Holly bought its first hatchery, and in 1957 the company added two processing plants.[8]

Fred Lovette was, in short, helping his company become vertically integrated. By incorporating previously independent facets of the industry (such as hatching, feed delivery, and processing) into a single business enterprise, Holly Farms was able to standardize and guarantee that chicks, feed, medicine, and advice were received by Holly Farms growers in time to ensure that full-grown broilers were consistently delivered to Holly Farms processing plants.

A certain amount of horizontal integration—"getting bigger"—not only was an inherent part of this vertical integration, but also was necessary to survive the periodic price fluctuations that characterized the industry. The "great merger" of 1961, which established Holly as *the* poultry company in Wilkes County and an industry leader, was part of this broader process of integration and industry consolidation. The merger itself, in which Fred Lovette successfully incorporated sixteen independent companies into Holly Farms, was the most significant merger financed by private capital in the history of the poultry industry. It culminated the vertical integration of Holly Farms while simultaneously jump-starting the process of horizontal integration. Subsequent mergers and acquisitions were less about increasing efficiency through integration and more about getting bigger and gaining market share.

For the Wilkes-based companies involved in the merger, there was little choice but to integrate. Independent firms were in a period of disastrously low poultry prices, and feed companies and hatcheries were in serious trouble. Most of the companies were already doing business together, but they lacked the coordination and size necessary to help each other weather periodic low prices—consequently, problems in one area reverberated throughout the entire supply and production chain. This situation, combined with overall trends within the industry, sent a clear message to small, independent, firms: merge or die.[9]

The 1961 merger included three hatcheries, a feed mill with grain purchasing and hauling subsidiaries, five broiler contractors, a breeder-flock company, processing facilities, and three related organizations: Lovette Poultry, Pierce Poultry, and North Wilkesboro Ice and Fuel. The much larger Holly Farms, still controlled by the Lovette family, was now a major force within the industry. The company had 1,100 growers, 850 employees, the nation's largest processing plant, and a production capacity of around 650,000 birds a week. Holly's dressed poultry sales, which reached much of the southeastern United States, were expected to top $20 million in the year following the merger. The company primarily produced fresh (ice-packed) chicken but planned to diversify into branded, prepackaged products as well as "convenience foods" (if the idea ever caught on with consumers). Wilkes County had arrived.[10]

During the next quarter century, from 1961 until the late 1980s, Holly Farms would grow into one of the largest poultry producers in the world. This dramatic growth was driven in part by factors outside the company's control (such as the growing demand for chicken). But Holly made a number of key innovations, including an early push into further-processed, value-added poultry products. Only three years after the merger, in

1964, Holly first tested the chill pack, or what would become known as Holly Pak. Before this advance, whole chickens were typically packed in ice, sent to market, cut by the butcher, and then displayed. The chicken not only absorbed a lot of moisture from the ice, but also went through a series of temperature changes during transportation that led to earlier spoilage. There was also the problem of selection. Because the butcher acquired the whole chicken, he had to sell two breasts, two legs, two wings, a back, and the rest of the bird before anything spoiled. By contrast, Holly Pak kept chicken at a constant temperature just above its freezing point (twenty-eight degrees), and also cut, packaged, and shipped the chicken in parts according to what individual customers desired (more legs than breasts, for example). This magical temperature of twenty-eight degrees meant that "fresh" chicken now had a much longer shelf life.[11]

Holly Pak revolutionized the company and the industry. By 1968, Holly Farms was at the top of the industry, having acquired an additional twenty-two companies and boasting sales of over $100 million. In addition to its Wilkesboro plant—the largest in the world—the company had four other plants that together processed and shipped up to 200,000 chickens a day to twenty-two different states. Holly's success fueled even bigger plans. By the mid-to-late 1960s, it was apparent that the industry's future was in further-processed, value-added convenience foods. The more value poultry companies added to the raw chicken, the less vulnerable they were to the price swings affecting basic commodities. Societal changes, including women's entrance into the paid labor force, helped create a market for these further-processed food products. Consequently, by 1968, Holly had plans to construct additional processing plants that would be devoted entirely to further processing, cooked poultry products, and eventually even whole meals. The company also wanted to start its own restaurants and partner with Safeway supermarkets to sell Holly Farms Fried Chicken directly to the consumer.

Plans require money, however, and in September of 1968 the Federal Compress and Warehouse of Memphis bought Holly Farms in a friendly acquisition. Federal was in the business of converting basic agricultural products such as wheat, grains, fruit, and poultry into consumer products. It had a feed mill and flour milling operation, but its main business was storing cotton. Federal was cash rich, looking for an investment, and had agreed to bankroll Holly's expansion.

With financing from Federal, Holly was able to move aggressively into branded chicken products during the 1970s. As chicken became no longer just chicken, and as Tyson, Perdue, and Holly Farms began to develop and patent an array of processing techniques and poultry products, the poultry giants began to realize the importance of branding and self-promotion. At the beginning of the 1970s Holly had virtually no name recognition. By the end of the decade, however, in markets where the company advertised, between 85 and 95 percent of consumers recognized the Holly chicken name (a better result than for virtually any other product sold in grocery stores).[12]

Holly's growth during the 1960s and 1970s was remarkable. When Fred Lovette orchestrated the great merger of 1961, the company processed 650,000 birds a year. In 1979, the company was marketing 5.4 million chickens *a week*. The company now had over one hundred Holly Farms Chicken Restaurants in five states to go along with its nine processing plants. The Wilkesboro plant was still the largest fresh chicken plant in the world, churning out 250,000 birds every eight-hour shift. Indeed, 14 percent of all fresh chicken bought in the United States now carried the Holly label. The effect of Holly's expansion on North Carolina and Wilkes County was no less dramatic. By the late 1970s, one in seven working adults in Wilkes County was employed by Holly Farms. In 1961, before the rapid period of growth, Holly employed 1,500 people in North Carolina, including 850 in Wilkes County. In contrast, by 1979, the poultry giant employed over 3,700 people, with some 2,400 in Wilkes County (with an additional 1,300 semi-independent growers supplying the company with its broilers).[13] Holly Farms was a major force in Wilkes County, in the poultry industry, and on American supermarket shelves.

The Tyson Takeover

Holly Farms rose to the top of the chicken world with amazing speed. Record year followed record year as Americans ate more and more chicken. The year 1984 broke the mold, 1985 was nearly as good, and 1987 was the best year financially in company history. By the mid-1980s, Holly was selling over $860 million worth of poultry products in forty-two states and employing more than ten thousand people. The poultry giant was the country's largest producer, processor, and marketer of branded fresh broiler chicken.[14]

Nor was the Wilkes-based company alone. The corporations that dominated the industry in the mid-1980s succeeded in remarkably similar ways: vertical integration, mergers, and expansion into further-processed poultry products.

In an industry characterized by low profit margins, the drive to expand and increase market share is a never-ending one. A 2 percent profit on $1 million in sales is small change, whereas a 2 percent profit on $15 billion is something else entirely. As a result, industry consolidation through mergers and acquisitions continued through the 1980s but on a much grander scale than before, eventually consuming even Holly itself. The pride of Wilkes County fell victim to Arkansas-based Tyson Foods and the very forces that had made it an industry leader in the first place.

The Tyson-Holly merger was one of the nastiest and longest takeovers in corporate history. The first, second, and fifth largest poultry companies in the world were locked in a fight to the finish. Unlike the 1961 merger, this was not about forming an integrated company by consolidating independent firms into a single enterprise. The Tyson-Holly merger was about one large, national, and fully integrated corporation taking over another. It unified two of the largest agribusiness corporations in the world. As fewer and fewer firms controlled a larger and larger share of the industry, the very essence of capitalism—competition—was eroded on the national level and virtually eradicated locally. Growers and workers increasingly found their communities dominated by a single integrator with few local ties or attachments. And consumers found that the chicken they ate—its cost, form, and health—was controlled by a handful of exceptionally large corporations.

In this sense, Tyson's takeover of Holly was a culminating moment in the shift from an industry characterized by local flavor to one defined by corporate blandness. To be sure, Holly was itself a full-fledged corporation by the 1980s. Its headquarters were in Memphis (after the Federal merger in 1968), and it produced and sold chicken throughout much of the United States. But growers and workers could nonetheless maintain the belief (or hope) that Holly was local, homegrown, and "one of us." The heart of its poultry operations was still based in Wilkesboro, and the Lovette family remained central to the company's operation. Paternalistic vestiges of the "good old days" were still present. Personal relations between growers, workers, and the highest levels of management remained a key part of Holly's day-to-day operations.

This intimacy would be shattered in 1988, symbolically at first with the death of Fred Lovette, and then definitively with the merger itself. Holly Farms, Perdue, and Tyson would all go through something of an identity crisis when their aging patriarchs—Fred Lovette, Frank Perdue, and Don Tyson—loosened their grips during the 1990s. Each had a populist, down-home, rags-to-riches quality that tied workers and growers (and in the case of Perdue, even consumers) to their companies long after they had become corporate Goliaths. In the case of Holly, Fred's death stimulated the merger itself by removing the last impediment to the Tyson takeover.

In the end, however, Fred Lovette's death only signaled what insiders at Holly must have known by late 1987: Holly was vulnerable. Even in 1987, Holly's best year ever, the company was in trouble. Its more than one hundred wholly owned restaurants proved a disaster. They closed during the first half of the 1980s, thus ending the company's foray into retail. Then, beginning in 1987, the industry as a whole went through a period of overproduction as growers became more efficient, which drove chicken prices down. For those integrators, such as Tyson, that had completed the transition to further-processed, cooked chicken, these swings were difficult but not disastrous. But despite important innovations during the 1970s, Holly had never completed the shift into further processing. It was an industry leader in branding, which made it an attractive acquisition. Yet Holly Farms remained, more than any other large integrator, a producer of fresh chicken sold in supermarkets. As a result, the company was more vulnerable to price swings.

By 1988, Holly was forced to respond to falling profits, and it did so in two ways. First, it reduced the number of growers from about 1,300 in 1979 to fewer than 500 by 1988. (This change was driven in part by the company's desire to work with a smaller number of large growers, as opposed to many small growers, but it also reflects a period of austerity.) Second, and more important, in 1986, just as Holly was closing its last restaurant, it shifted its focus toward prepared foods. In late 1987, Holly completed a $20-million roast chicken plant right next to its existing processing facilities in Wilkesboro. The plant, which became Holly's cooked products division, produced roasted chicken in original, Cajun, and barbeque flavors. The idea was to get the consumer out of fast-food restaurants and back into the grocery store. At about the same time, Holly bought Weaver Inc., the leading retail brand of frozen chicken sold in the Northeast. Because Holly sold virtually none of its chicken in retail frozen food cases, this acquisition not only expanded its geographic reach, but also accelerated its push into further processing.[15] It seemed like an ideal match.

But it was too little, too late. Aside from supply problems, the chicken market proved sluggish because of publicity surrounding *Salmonella*. Holly was the number one brand of fresh chicken sold in supermarkets, and that was precisely the problem. The company's 1988 annual report laid it out in blunt terms: "We lost money on chicken—our worst year in history. And the poor performance was mostly our fault." The company was producing too much chicken, too expensively. Increased help from its line of further-processed products could not change the simple fact that the company's mainstay—fresh chicken—was an economic loser. Worse yet, Holly's solution, Oven-Roasted Chicken, had become a problem. Poor marketing resulted in sluggish sales.[16] Holly had recognized the need to get more serious about further processing, but it had chosen the wrong product to promote.

During the next year, Holly Farms concentrated on becoming more efficient by laying off workers, cutting growers, and consolidating operations. Ironically, however, this downsizing made the company not only more vulnerable to takeover, but also more attractive to industry leaders. Don Tyson had approached Fred Lovette a number of times about a merger. The two went way back, and early merger talks were friendly. Fred had even shared the Holly Pak innovation with Don in the 1970s. In October of 1988, however, with Lovette having just passed away, Tyson dispensed with the pleasantries and made an uninvited bid to take over Holly Farms Corporation. The Arkansas behemoth offered $45, along with a quarter share of its Class A common stock, for every share of Holly Farms, which amounted to close to $1 billion.[17] At the time, Holly was the fifth largest producer in the country and the leader in branded fresh chicken.

From Tyson's perspective, the merger seemed ideal. With its acquisition of Valmac Industries and Lane Processing in 1986, Tyson was already the leading poultry producer in the United States.[18] By acquiring Holly, Tyson would control over 28 percent of the country's chicken supply (up from 18 percent). More important, Tyson would be bigger *and* better. The merger, which would combine the industry leader in further-processing (Tyson) with the world's largest producer of branded fresh chicken (Holly), was a textbook example of corporate synergy. Tyson Foods would take a giant step into fresh-branded chicken, gain market access to the southeastern United States, and achieve unprecedented control over America's favorite meat.

Holly immediately went on the defensive, rejecting Tyson's offer without discussion.[19] This would not be a friendly merger.

The central problem for Holly, aside from the low bid price, was the sharp difference in management style. Don Tyson was a risk-taking maverick who was willing to take on massive debt in order to make Tyson Foods the king of chicken. By contrast, Lee Taylor, the forty-six-year-old Princeton-educated head of Holly Farms, was much more conservative and low-key. Few doubted Taylor's resolve, however, or his willingness to attract other suitors in order to stave off Tyson's challenge.[20]

The Tyson–Holly affair was only one among a series of mergers that were rocking the financial and food world during the late 1980s. At the exact moment when Tyson was trying to acquire Holly, the RJR Nabisco deal went through, the Philip Morris Companies offered $11.1 billion for Kraft, and Grand Metropolitan PLC, a British beverage and food giant, was attempting to take over Pillsbury at a cost of $5.23 billion. Part of what was fueling merger mania was politics. Ronald Reagan was leaving office, and no one expected the next president—Republican or Democrat—to be as merger-friendly. Part of it was also cyclical. Food companies, especially ones with familiar brand names, are attractive during periods of market uncertainty.[21]

Tyson next proposed $52 a share, an offer that Holly also promptly rejected.[22] In fact, four days later, Holly dropped a bombshell. It agreed to merge with ConAgra Inc., the second largest poultry processor in the country (and owner of Butterball and Country Pride brands), for about $55 a share, or $957 million. Refusing even to talk with Tyson, Holly had signed on with ConAgra after almost no discussion. Tyson had been derailed, at least temporarily. If ConAgra acquired Holly, it would replace Tyson as the world leader in chicken.

Tyson went on the offensive, making a $57-a-share merger proposal to the Holly Farms board. Holly immediately rejected

Tyson's third offer, insisting that ConAgra's offer not only was far superior, but had already been accepted. Tyson quickly turned to the Delaware Chancery Court to invalidate two rather suspicious provisions of Holly Farms's merger agreement with ConAgra. The first was a "lockup" condition that gave Con-Agra the option of purchasing Holly's poultry assets regardless of whether the merger went through. This provision was clearly designed to stop Tyson's bid. Why would Tyson want Holly if the company had no poultry? The second provision was also aimed at Tyson: Holly would pay ConAgra a hefty termination fee if the merger fell through and Holly was acquired by a third party. Tyson argued that in rejecting its offer the Holly board had not pursued its stockholders' right to the best price and had done so simply because it was unfriendly toward Tyson's management.[23]

On the eve of the New Year, Judge Maurice Hartnett of the Chancery Court in Delaware blocked Holly from merging with ConAgra, an action that prompted Tyson to raise its bid to $60 a share. Judge Hartnett ruled that provisions of the agreement did not serve the best interests of Holly stockholders because it did not give them the opportunity to consider Tyson's offer. Holly acquiesced and put the company up for auction. What could be more fair? And why not? Holly's stock price had been in the low forties when Tyson had made its first overtures in October. It was now January, $60 a share was on the table, and the industry's two giants appeared willing to do just about anything to acquire Holly. Was $65 a share out of the question?[24]

With ConAgra ignoring the request for new bids, Tyson raised its offer to $63.50 just before auction day. A lot was riding on the deal. Without additional expansion, Tyson would reach maximum production by about April. Unable to meet additional demand, the company would then begin to lose market share. Expanding from the ground up was not Tyson's forte. Thirty of its thirty-two plants had been obtained through mergers and acquisitions.[25]

On February 7, Holly announced that, after reviewing bids from its two suitors, the board of directors had unanimously confirmed its earlier merger agreement with ConAgra. Yes, ConAgra's bid was lower, but the company expressed doubts about Tyson's ability to consummate its bid because of the provisions associated with the merger agreement with ConAgra. On April 17, however, the merger bid from ConAgra was soundly rejected by Holly shareholders.[26]

Everything was in limbo until Holly dropped yet another bombshell on May 22. It had entered into a "definitive" agreement to merge with ConAgra, in an exchange of stock that valued Holly's at about $75 a share. Like the earlier agreement, Holly would pay ConAgra a $15-million fee and up to $10 million in expenses if the deal fell through. Tyson merely pointed out the obvious. Nothing was "definitive." Tyson raised its cash offer to $70 a share and removed the condition that Holly end the lockup agreement with ConAgra. Tyson's condition-free offer, which was in cash as opposed to a stock swap, was now more attractive. Tyson also asked for $500 million in damages, accusing Holly of wrongly favoring ConAgra. Tyson was now suing the very company it was trying to take over. Holly had little choice but to announce that it had entered into negotiations to end its merger agreement with ConAgra and to accept the acquisition offer from Tyson. The only remaining question was how much ConAgra would be compensated. Turning to the Delaware court once again, Tyson was ordered to pay ConAgra $50 million for its trouble.[27]

By acquiring Holly, Tyson solidified its position as the top producer of chicken in the United States. Tyson already controlled about 60 percent of the wholesale business that supplied fast-food restaurants and other institutions.[28] The acquisition of Holly gave Tyson a dominant presence and name in the retail business, which would allow the company to sell the products it had developed for restaurants directly to the consumer. Holly was a force in grocery stores, controlling 40 percent of the fresh-poultry market in New York City and Washington, D.C.[29] But Holly was precious to Tyson not simply because it produced a lot of chicken. Tyson fought fiercely and paid dearly for the recognition of Holly Farms's name by consumers in the nation's largest markets. It was the industry's first brand war.

From a number of perspectives the merger appeared to have no downside. Lawyers and consultants got rich. Tyson became bigger and thus better able to compete in the increasingly global poultry industry. To be sure, Holly no longer existed. But it was now part of the largest poultry company in the world. Moreover, as one of the key figures pointed out, "We hated to see Holly get taken over by Tyson. We had built this company and it was sad to watch. But man, we cried all the way to the bank. When it all started the stock price was in the forties. We sold for seventy."[30] R. Lee Taylor, Holly president and chief executive, made over $15 million from the deal. Blake Lovette made a fortune and had no trouble finding employment. After leaving Tyson, he wound up as an executive vice president at Perdue and then as president of none other than ConAgra Poultry Co. As one of the principals put it, "The merger produced more millionaires in one day than Wilkes County had in all its history." Bank deposits at the local First Union increased by 20 percent in the months following the merger, leading one bank official to speculate: "My personal feeling is that the Tyson buyout of Holly Farms brought about

$100 million into the county."[31] For its part, ConAgra would walk off with $50 million, more than enough to cover its expenses and make new investments of its own.

If one were to believe the media and the "movers and shakers" within the industry, a handful of executives at Tyson, Holly, and ConAgra had played a high-stakes game that was intriguing, but had few consequences for ordinary people. Shareholders occasionally entered the picture, but workers, growers, and consumers were completely absent. This is fitting. The folks who produce, process, and consume chicken for Holly, Tyson, and ConAgra were and are not only completely irrelevant to the merger; under the current system, they are oddly incidental to food itself. Product development, marketing, packaging, and distribution remain the foundations of the contemporary food industry. Production and consumption, two of the most basic features of food, are strangely irrelevant to our industrial food system. In Chapter 4, I suggest that although producers may have been irrelevant to the merger, the merger was not irrelevant to them.

IV

The Right to Work

On July 18, 1989, Don Tyson and Blake Lovette shook hands in a public ceremony celebrating the merger and Tyson's first day at the helm. The future was bright. Wilkesboro would continue to be the home of fresh chicken, now under the direction of the world's largest poultry company. No jobs would be lost, and continued expansion could be expected. Don Tyson wasted no time in trying to win over the workers with his down-home antiunion philosophy: "Why should you and I, as individuals, have to have somebody work between us? It's like hiring a lawyer, and both of us paying him, when we could have thrown him out the window. In the last few years, of the companies that came with us, five plants that were union voted them out, where they belong." Tyson Foods, according to Don Tyson, was "pro-people." As one worker commented wryly: "It's got to be better [under Tyson]. It can't be any worse."[1]

It was clear from the beginning, however, that more than

the company logo was going to change. Holly management was the first to go, starting at the top. Blake Lovette resigned from the company in August 1990, and decision-making was formally moved from Wilkesboro to Arkansas.[2] Jobs in marketing and sales, along with those in accounting, human resources, aviation, commodities, and transportation, were slowly cut or moved to Arkansas. Eventually, the Holly Farms headquarters was shut down and sold, resulting in the loss of the last 175 jobs on-site. This means that by 1992, or about three years after Don Tyson announced that no jobs would be lost, there were close to two hundred fewer jobs in Wilkesboro.[3] Worse yet, these were among the best jobs and had been supporting a significant portion of the town's middle class. Wilkesboro's class structure became even more polarized. As local banks overflowed with merger-related deposits from high-level executives, the majority of the county's residents struggled for their economic survival.

The effect of Tyson's takeover on administrative and managerial jobs may have been swift and not entirely unexpected, but its broader impact on workers and farmers has been more subtle and difficult to judge. Most agree that "it got worse" after Tyson took over, but people also admit that the situation was headed in the same direction under Holly. Indeed, Tyson inherited something of a mess in Wilkesboro. Holly's attempt to cut costs may have made the company leaner and more attractive to corporate raiders, but it also had made life more difficult for workers and growers, who struggled even during the best of times. Further, during its battle with Tyson and ConAgra, Holly management had been engaged in two bitter disputes with workers and farmers. The first was a relatively straightforward union battle between the company and a group of truck drivers, a conflict that would become increasingly intense once Tyson took over and got into the mix. The second involved Holly's abrupt termination of over fifty of its most loyal and long-term growers.

An Antiunion Campaign

At the end of 1988, just as the Tyson–ConAgra battle was building steam, Holly Farms implemented the first in a series of cost-cutting measures designed to help the company bounce back from a disastrous fiscal year. Holly eliminated waiting-time pay and reduced layover pay for its long-haul truck drivers—the people who move chicken from processing plant to market. To make this new policy more palatable, Holly informed the truckers that they could drive more miles in order to compensate for the loss in income. In retrospect, Holly miscalculated. Who would have thought that the policy would strike such a nerve, leading to a ten-year struggle with the Teamsters and stimulating unionizing throughout the company's operations? As a striker pointed out, one thing that Holly learned from the experience was, "Don't mess with truckers."[4]

The Teamsters immediately entered the scene and pointed out the obvious. Working more for the same amount of pay was hardly a good deal. In fact, it seemed a lot like a pay cut. This common sense was attacked by editors of the local newspaper who couldn't understand why people in Wilkes County did not like to work more for less: "Union statements of that sort illustrate how the Teamsters are in opposition to the work ethic that's a key part of Wilkes County heritage." But the truckers would have nothing to do with the romantic notions of the press. In March, in a state that routinely vies for the least unionized in the country, the long haulers went against local wisdom and voted 211 to 70 in favor of union representation.[5]

The mobilization proved something of a spark for workers throughout the Holly complex. Ninety percent of over two hundred hatchery workers quickly signed cards to form a union chapter of their own. Although their complaints differed from those of the truckers, Holly's cost cutting was behind all of them. For example, beginning in the fall of 1988, at about the time Tyson made its first bid for Holly, some hatchery workers had been assigned jobs that had previously been done by two workers. In addition, prior to that period workers had been paid by the hour regardless of volume; then the company implemented a system of piecework and reduced workers' hours. Holly's response was also a familiar one. Officials explained that increased mechanization was now allowing one person to do what had previously been two jobs; and besides, folks who worked in the hatchery, chicken catching, and much of live production were "agricultural workers" and hence not allowed to unionize under the law.[6]

Other workers got into the fray, started organizing, and in at least one case held a union election. That this was going on during a period of austerity, when jobs were scarce, testifies to the importance of the issue for workers. Holly countered with an aggressive antiunion campaign. Three long-haul truck drivers were arrested for trespassing while they distributed union materials in the company parking lot. Four other employees were discharged for soliciting union support while on the job. Other employees were threatened with dismissal if they selected the union as their bargaining agent. And the company even threatened to contract out the entire trucking operation if the workers went union.[7]

Once Tyson took over, the antiunion campaign shifted

into high gear. In a deliberate effort to get the union out, Tyson integrated Holly's unionized division with its own larger, nonunionized transportation division. Tyson then argued that another vote was needed in order for the union to be recognized as the bargaining agent for the workers—because the unionized Holly truckers no longer constituted a majority of the total truckers. Tyson also forced Holly truckers to sign a card saying they would work for Tyson, thereby implicitly accepting their nonunion status, lower wages, and inferior working conditions. The workers' response was unequivocal. On October 1, 1989, the truckers, firmly supported by the Teamsters, went on strike and established a twenty-four-hour picket line at the entrance to the Wilkesboro plant.[8] The picket was to last three years. The struggle lasted ten.

During the entire ordeal, neither Holly nor Tyson won a significant legal battle. The workers and the union won ruling after ruling establishing that Holly and then Tyson engaged in a deliberate and sustained pattern of illegal antiunion activity. In March 1990, for example, the North Carolina Employment Security Commission ruled that twenty-four former Holly long-haul drivers should have received unemployment insurance benefits because they had good cause for rejecting jobs with Tyson. Holly–Tyson countered that because the workers had refused to sign the cards, they had in effect "quit" their jobs and were not entitled to such benefits. The judge was not fooled, pointing out that the workers rejected work that was unsuitable. "To earn as much or more, the drivers would have to have driven more miles, avoid[ed] accruing motel expenses which were not reimbursable, and remain[ed] away from families or home for a longer period of time." Worse yet, with Holly the drivers had already been driving the maximum allowable under the law. Under the Tyson plan, in order to maintain their salary, the truckers would have had to break federal law. As the judge pointed out, Tyson had violated the merger agreement with Holly as well as federal regulations by reducing the drivers' pay. Moreover, since no other employees had received pay cuts as a result of the merger, the judge couldn't help but wonder whether those particular drivers had been singled out because they were unionized and on strike.[9]

In March 1992, after numerous delays, Administrative Law Judge Robert Schwarzbart ordered Tyson to rehire the truckers.[10] Tyson immediately appealed, which put the decision on hold for another three years. Finally, in March of 1995, the U.S. Court of Appeals for the Fourth Circuit upheld the National Labor Relations Board (NLRB) rulings that Tyson had committed unfair labor practices. The NLRB, in fact, upheld nearly every ruling made by Judge Schwarzbart three years earlier, confirming that Tyson was guilty of unfair labor practices that were "serious, pervasive, numerous and calculated," and that involved the "highest levels of management."[11] By the time it got to the NLRB, the ruling involved three groups of Holly workers that had been involved with the Teamsters since December of 1988 (while Tyson's bid to purchase Holly was pending). In the case of the long-haul truck drivers, the court ordered Tyson to recognize the Teamsters as the truckers' bargaining agent, offer the workers their jobs back, and reimburse them for back pay. As for the other groups, the court ruled that Holly's harassment and threats against their union campaign had all been illegal.[12]

Almost a year later, in January 1996, twenty-seven truckers returned to work, now for Tyson. None were particularly happy with the conditions they found. Most of the other workers eligible for rehire had moved on, either finding other jobs in the area or leaving Wilkesboro altogether.[13]

These struggles were major victories for the Teamsters. The union had stood up for workers and persevered through an absurdly long struggle against one of the largest corporations in the world. The conflict cost the union well over $1 million, but the rulings set significant precedents. At the same time, it's unclear what the larger message was. Tyson affirmed that its well-worn strategy of "harass, stall, and appeal" might not always be legally effective, but it succeeded in draining union resources and making workers think twice about unionizing (or even complaining). As the industry consolidated and enormous companies like Tyson were put in charge, the playing field became so skewed that even a union with the history and resources of the Teamsters faced terrible odds. And for workers, even those who held all the legal cards, taking on a company the size of Tyson could mean the loss of an entire way of life:

> We had to go on strike and form a union when Holly cut our pay. They wanted more and more work for less and less pay. We just couldn't sustain it. So we organized. Tyson comes in and tries to play like our union didn't exist. So we had to fight. This I will believe until the day I die. We had to do it. Was it worth it? Hell no! Don't let anyone tell you it was. People lost houses, family. They had to move. And this was with all the help the Teamsters could give us. . . . This consumed some of us for ten years. We fought and we ended up with the same crappy job we had in the first place. Makes you wonder. When I look at what it did to me, my family, I can't help but think I should have just accepted what Holly

and Tyson were doing. But I couldn't have done
that either. They [Holly–Tyson] just had no respect
for us. I don't know if they do now, but I can at least
respect myself.[14]

The Growers' Predicament

In July 1988, Holly Farms informed fifty-four poultry growers
that their contracts would not be renewed after September.
The growers had received their last batch of chicks and would
be out of the business within weeks. The announcement,
which came at a time when Holly was trying to cut costs and
Tyson was preparing its bid, caught the growers completely by
surprise. In fact, all signals had seemed to point toward a
steady supply of chicks:

> We knew Holly was going through a tough time. So
> I kept asking my service rep if I was secure. He told
> me time and time again that I had nothing to worry
> about. We even got a letter from Holly saying they
> would stick with us. The worst part was that just
> before they cut us off they told us to make im-
> provements to our houses. We had to pay for these
> improvements, and many had to take out a loan. I
> hardly had the improvements done and he tells me
> this is my last batch [of chicks]. I had been growing
> chickens for Holly for twenty years. I had won
> awards as one of their best growers. I was still one
> of the best when they cut me off. They said we were
> too far away; that the truck had trouble getting up
> the mountain in winter. Bullshit.[15]

Most of the growers had been in the business for over ten years, and many had been growing for over twenty. They, as much as the Lovettes, had built Holly Farms. Some depended on chickens for the majority of their income, while others relied heavily on off-farm sources. Levels of debt also varied considerably. Some of the older growers had their debt paid off, while others owed remarkable sums of money.

None were hit harder than the Watson family. In 1981, Kenneth Goodman had approached the president of Holly Farms about getting his daughter and son-in-law, Sherl and Dennis Watson, into the chicken business.[16] Goodman's two houses had been supplying Holly with chickens since about 1970, and his father had also been in the business "before Holly was Holly."[17] Sherl Watson would be a third-generation grower.

The idea was simple. Sherl and Dennis Watson would take a twenty-year loan to cover the purchase of a 7.5-acre farm and the construction of four chicken houses. The loan was for $210,000, a substantial sum in the early 1980s for a couple who had only been married a year and had no resources. To guarantee the loan, Sherl's father put up his house as collateral. As Dennis recounted, "We didn't feel totally comfortable . . . him putting up his house like that. It was a great thing to do. We had nothing and it seemed like a sure thing. We knew it would be hard work. But we thought if we did everything we were supposed to, we could pay it off. And that's just it. We *did* everything we were supposed to. We had everything on the line. Our house, our farm, Sherl's Dad's house. Everything."[18] At the time, the president of Holly assured Goodman that his daughter would have a contract to raise chickens at least through the duration of the loan. This was the early 1980s, and the future was bright. The loan itself was from a bank, but the payments were deducted by Holly and then forwarded to the bank. The Watsons could not possibly have gotten the loan in the first place had Holly not facilitated it, thereby effectively assuring the bank that it would do business with the Watsons.[19]

According to all accounts, the Watsons were model growers and raised broilers for Holly without incident until 1987. They kept vigil over their chickens seven days a week, leaving the farm for only one night during an eight-year period. As Sherl put it: "We made the payments and lived. Nothing more. We just got by, but we were buying a small house and living on a farm. We were together, with our kids. . . . You can't put a price on that. That was the best thing about it, really. It let us be a family."[20]

In January 1988, Holly informed the Watsons that they had to install new feeders in their chicken houses. With Holly facilitating the arrangements, the Watsons took out two more loans, totaling nearly $20,000 and at 12.5 percent annual interest, to make the improvements. The loan was to be paid in twelve quarterly payments beginning in June 1988. "You've got to understand. It's not like we had finished paying off the original loan. That extra loan nearly killed us. But when you got four chicken houses and your whole life is on the line you can't really argue. So we made the improvements."[21]

Only eighteen months later, the Watsons were cut off:[22]

> Jerry, our serviceman, came over and told us to sit down. We had a good relationship with him. But it's never good when they tell you to sit down. Then he reads this letter. You know something? He wouldn't even give us a copy of the letter. Just unbelievable. You would think they would have sent some company official. Nope. They just send poor

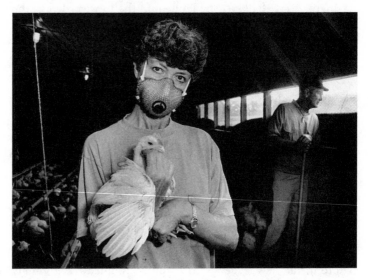

Poultry farmer (Photo by Earl Dotter)

Jerry. He reads it. Tells us we won't be getting chickens after October 1. It was like someone just punched me in the stomach. We didn't even know what to say. We knew we lost everything right then. We tried everything, but they wouldn't talk to us.[23]

The Watsons were in trouble. Completely dependent on chickens for their income, they immediately started working any job they could find in order to keep up with the payments. It was impossible. They quickly defaulted on their loans and lost everything. As Sherl Watson put it:

I couldn't even talk about the whole thing until three years ago. This was our dream and it fell apart.

We didn't just lose our income. We lost everything. It ruined my life for two to three years. Really.

Look. We were about ten years into a twenty-year loan. We couldn't file for bankruptcy. We're not that kind of people. Plus if we did my Dad would lose his house. Can you imagine? So we had to sell everything. I mean everything. The house. The farm. Every piece of farm equipment we owned. Anything of value. With that money we found a bank that made us a loan so we could pay off the other loan. We were paying for chicken houses, a farm, and a house we no longer worked or lived in. For years we did this. I'm not sure how we kept it together.

Dennis added:

We went in with nothing and came out with nothing. Actually, we were worse [off] than [when] we started. That's how I look at it. The worst thing is we lost ten years of our lives. Ten years. Our most productive years were gone. We would have had everything paid off by now. This house here we still haven't paid off. You can't make up ten years. We will be paying for that the rest of our lives.[24]

The Watsons sued Holly Farms, arguing that the loan for the feeders, which Holly collected and forwarded to the bank, was effectively with Holly; or, at the very least, the loan was contingent on Holly collecting the birds. They sued for loss of income, loss of farm, loss of the amount paid for (now useless)

Poultry farmer at work (Photo by Earl Dotter)

equipment, loss of credit and business reputation, and other damages in excess of $10,000. They alleged unfair or deceptive trade practices, as well as emotional distress and mental anguish. According to the suit, Holly had been reckless in requiring the Watsons to purchase the feeders without disclosing that the contract was about to be terminated, or that termination was a reasonable possibility. In short, Holly's actions were "extreme and outrageous." Holly denied it all, settling with the Watsons out of court.[25] The Watsons could not reveal the exact amount they received. At best, it allowed them to pay for feeders they never should have installed.

Few growers were as vulnerable as the Watsons at the time of the termination. But they were all in trouble and quickly formed the Mountain Poultry Growers Association, an organization that included the growers who lived in Ashe,

Broiler house, mid-1970s (Courtesy USDA)

Watauga, and Alleghany counties and who had been cut off. As one grower wryly noted:

> Once we formed the organization, we were dead. We had been dealing with Holly people for all our lives and suddenly we couldn't even get them on the phone. Don Tyson never gave us the time of day. The companies don't like it if you organize. We knew this. We formed the organization out of desperation. We thought if we could get publicity maybe we could pressure Holly. That worked a little bit, but then the merger took over and no one cared. The other thing we wanted to do was get a contract with someone else. This was a long shot because we were too far away from everyone except Holly. It may seem silly now, but we wanted to show that we had a group of experienced growers who

[together] could raise a lot of birds. For this, we
needed to organize.[26]

In the initial months after the growers were cut off, a
number of possibilities emerged. At the beginning of 1989,
when Don Tyson was still trying to woo Holly, he told the
growers that if he bought Holly he would keep them on. The
Mountain Growers were no fools. Don Tyson would say just
about anything in order to look better than ConAgra. (And, in
fact, on Tyson's first day in Wilkesboro, while he was reassur-
ing workers their jobs would be protected, Holly confirmed
that the decision to cut off the Mountain Growers was final.)[27]
As a result, the Mountain Growers explored a variety of
other avenues, including switching to raising turkeys, in the
months following their termination.[28] But viable options were
few and far between once Holly had pulled out. As one grower
pointed out, "There weren't a lot of options around here *before*
Holly. And it was never a picnic with Holly. Most of us still had
to struggle to piece together a living. But we got by. When
Holly pulled out it wasn't simply that we lost a source of income, as if we had lost a job. . . . We lost income and still had
to pay off debt. And the fact is you can pretty much do only
one thing with chicken houses. Without chickens, the houses
are useless."[29]

It took almost three years before any of the growers were
able to find a use for their chicken houses, and then only
briefly. In the summer of 1991, a group of fewer than ten
growers began raising chickens for Morganton-based Breeden
Holdings. Breeden had too many chicks and not enough growers, so it placed 29,000 birds with growers like Joe Brown, a
local farmer who had been with Holly for almost thirty years.
Like many Mountain Growers, Brown had cattle on what had

been a full-time farm until Holly pulled out. Over the years poultry had provided a fairly consistent income that was a nice complement to cattle, the price of which tended to fluctuate considerably from year to year. Together, cows and chickens allowed Brown to stay on the farm full-time. Once Holly stopped sending him chicks, however, it became impossible to maintain even the cattle, and Brown did what many others in the area were doing. He provided the land and half of the working capital for a Christmas-tree business, with a local tree nursery providing the rest of the capital and organizing the operation. The work of planting, caring for, and then harvesting the Christmas trees is done by Mexican immigrants while Brown works full-time at a leather goods company.[30]

Unfortunately, the relationship with Breeden was not only limited to a handful of growers, but also failed to last past the summer. Brown and the others were back out of the poultry business almost as soon as they had returned. Morganton, it turned out, was just too far to travel.

About five years after Holly–Tyson pulled out of the mountain counties, the growers embarked on their most interesting and ambitious experiment. In February 1994, a Kentucky-based processor of free-range chicken, Wilson Fields Farms, met with growers. At the time, Wilson was processing only 20,000 birds a week, mostly for upscale restaurants in New Orleans. By contrast, the Mountain Growers had been raising about 1.5 million birds per cycle per year under Holly. This was not a large-scale solution that would help the majority of growers. Nevertheless, Wilson Foods was a welcome arrival, both symbolically and economically.[31]

The project started with a handful of growers, at considerable cost, retrofitting their houses to make them free-range friendly. Wilson Fields then delivered several batches of chicks, the necessary feed, and collected the broilers. According to one grower, "Everything was going great. Changing the houses was a bit of trouble. I did a bunch of stuff and built this fence around the barn so the chickens could walk around. You know, they were supposed to 'range,' like they are at Disneyland or something. The funny thing is the damn things never left the barn. But that's what Wilson wanted, so I did it. And it was all worth it because in the long run it would be more profitable."[32]

Then rumors that Wilson was struggling began to circulate:

> Things were going great until the middle of the third or fourth batch. Then we heard Wilson was having problems. Then Wilson stopped bringing feed. At first we waited, but then things got desperate. The chickens started getting hungry and needed food. We couldn't afford to feed chickens we weren't going to sell. You get the feed on credit from the company that buys the chicks. Besides, chickens aren't pets. We're not feeding 25,000 chicks if we can't sell them. This is a business. *Oh,* but these people from Washington go nuts. They come down here and start picketing. They kept using this term. Damn. I can't remember it. [I interject: "animal cruelty"?] Exactly. Can you believe it? They said we were being cruel to chickens. We're raising them to be processed into nuggets so these people can eat them and they say we are being cruel.[33]

People for the Ethical Treatment of Animals (PETA) had learned of the growers' (and chickens') dilemma and were demanding that the birds be killed immediately to end their suffering. The chicks were not only hungry, but since they had not been debeaked they had also begun pecking each other. In response, two of the growers took out criminal summonses charging animal cruelty against the president of Wilson Fields. Sensing a public relations nightmare for the industry, the National Broiler Council stepped in and brokered a deal with Atlanta-based Gold Kist, one of the largest integrators in the country, to take the birds to its Athens plant. As one grower noted: "That got the idiots from Washington out of here, but it didn't help us or the chickens much. The chickens died on their way to Georgia. You can't transport chickens three hundred miles, especially weak ones. We were left with empty houses that we had just [retrofitted] for Wilson. What a disaster. I got a sense of humor about the whole thing now, but it wasn't real funny at the time."[34]

There is nothing funny about industry consolidation, mergers and acquisitions, or corporate takeovers. Most of us read or hear about these events with only casual interest. Rarely do we recognize that these corporate goings-on can have profound implications for all of us. When I began working at the Tyson plant in Wilkesboro, the merger was over a decade old. For many of the long-time workers, however, it remained a pivotal moment.

> I have been working here for over sixteen years. This has never been easy work. But under Holly it was decent work. . . . Now managers are under extreme pressure to produce and therefore the workers are, too. . . . They started going more and more into further-processed products. See, this plant used to be all fresh chicken. [Now] we are cutting it into little pieces, and doing all sorts of stuff to it,

even cooking it. The work in the past was just as physical, but now it is more repetitive. It's not just the speed. Sure it's much, much faster. But it is also the work. It's more intricate. It's physical and delicate at the same time. The worst thing, though, is that every time they change the product they see it as an opportunity to increase the workload. . . . The plant is more productive now than in the past. Much more productive. But it is not just because of better machines. It's because the workers work much harder. This is why you see so many Mexicans. . . . Managers say the Mexicans work harder. I guess it's true. But it's because they just got here and don't know what it used to be like.[35]

As power within the industry has become increasingly concentrated, workers, farmers, and consumers—those with the greatest stake in our food system—have been relegated to the margins. We have been on the outside looking in as a handful of corporate giants have transformed the basic terms under which Americans farm, work, and even eat.

II
A New Worker

V
Getting Here

Born in Chihuahua, Mexico, Amador Anchondo-Rascon first tried to cross into the United States in 1979. After a ten-day journey on foot, he was captured and returned to Mexico. He waited, crossed again, and by the age of twenty was an undocumented worker in the fields of Florida. After marrying an American and becoming a legal resident, Amador was drawn to Tennessee by the possibility of better employment. His journey took him first to the hill country of McMinnville, where he worked in agricultural nurseries, and then west to Shelbyville, where a Tyson Foods' processing plant promised even better wages.[1]

Amador was something of a pioneer. When he first arrived in Shelbyville in 1989 there were fewer than one hundred Latinos living in the entire town. Drawn by the Tyson plant, the Latino population was to grow to over 2,300 by the year 2000, or to 15 percent of the town's total population (not including those uncounted by the census). Amador was also ambitious. In 1995 he opened a small grocery, Los Tres Hermanos, named after his three sons. Immigrants, most of whom worked at the Tyson plant, gathered there, where, according to Amador, "they felt like they were in Mexico again." Before long Amador was both prosperous and popular. With basic English skills and good connections, he served as a cultural liaison for new arrivals as they tried to get a job at the plant, purchase cars, rent housing, or deal with the police and courts.[2]

Amador's success is exceptional among Mexican immigrants to the United States. Born one of eleven children from a poor farming family, he was living the American Dream about fifteen years after he arrived illegally.[3] Nevertheless, the broad outlines of his journey are common enough to anyone familiar with Latin-American immigration into America's heartland. Like many Mexicans arriving in the 1980s and 1990s, Amador found his first job in a labor market that was rapidly becoming saturated with immigrants. He chose the fields of Florida but could have just as easily ended up in Texas or California. Amador's subsequent move to Tennessee was also quite typical, in both its timing and direction.

Indeed, what started in the 1980s as a trickle of immigration to the South became, by the 1990s, a powerful torrent. Between 1990 and 2000 the states with the fastest growing "Hispanic" populations were all located in the South. The region as a whole received over half a million Latinos just during the 1990s, in effect tripling the Latino population throughout the region.[4] North Carolina experienced a fourfold increase in Latino residents; Arkansas wasn't far behind; Georgia and Tennessee each had around a threefold jump; and South Carolina and Alabama experienced a doubling of their Latino populations during the decade.[5] Although the causes bringing folks like Amador were numerous—including an economic crisis in Mexico, changing immigration laws, the saturation of the low-wage labor markets in California, and an economic boom in the U.S. South—the role of chicken cannot be underestimated. Half of all poultry processing is concentrated in Alabama, Arkansas, Georgia, and North Carolina, with Arkansas, Georgia, and North Carolina having the fastest growing Hispanic populations in the United States, while Alabama is among the top ten.

Clearly, other forms of employment attracted immigrants to the South, especially into urban areas. The growing high-tech, biomedical research, and banking industries have expanded service economies and the demand for low-wage labor in places like Atlanta, Austin, Raleigh-Durham, Charlotte, and Birmingham. Broader patterns of economic growth have also attracted Latin Americans to major cities like Memphis, Greensboro, Richmond, Charleston, and Huntsville.[6] As these cities grew, both economically and numerically, many working-class blacks and whites moved up the economic ladder and left low-wage jobs to Latin Americans and other foreign immigrants.

Even in nonmetropolitan areas, the immigration of Latin Americans cannot be explained completely by jobs in the chicken industry. Auto plants now dot Tennessee, Kentucky, Alabama, Mississippi, and South Carolina. Garment, textile, carpet, and furniture industries are also crucial sources of industrial employment outside of major urban centers. Even agricultural employment encompasses far more than food processing. Latin Americans are now closely identified with the planting, picking, and packing of fruits, vegetables, tobacco, shrubs, and even Christmas trees. In North Carolina, 90 percent of farmworkers are Hispanic.[7] The overwhelming presence of Latin Americans in agriculture has even led some to wonder: "Do Americans produce any of their own food?"

Latin Americans did arrive in remarkable numbers to work in poultry, however, and the industry has served to both concentrate and keep immigrants in regions of the South that had little previous experience with a foreign labor force. Unlike agriculture, poultry processing is not seasonal. Processing plants operate nearly all day, every day, and require a permanent labor force (with enough excess workers to replace those who cycle in and out of plants). Traditionally, poultry workers were drawn from the local population. But by 1990, as the southern economy expanded, local workers became increasingly unwilling to take on jobs in the plants, at least at prevailing wages.[8] Consequently, poultry companies turned to Latinos.

In short, poultry did what meatpacking has done for the Midwest.[9] It brought foreign workers like Amador into the heartland—that is, the smaller cities and towns that embody "the South." It is here—in places like Russellville, Alabama; Shelbyville, Tennessee; and Rogers, Arkansas—where Latin Americans have had their greatest influence. The arrival of 50,000 or even 100,000 Latin Americans into Atlanta has been significant and has required adjustment. The arrival of seven thousand Latin Americans into Duplin County, North Carolina—home to the largest turkey processing plant in the world—has been revolutionary. It not only has transformed schools, social services, housing, and local businesses; as we will see in Chapter 7, the "overnight" presence of a foreign, brown-skinned, Spanish-speaking population has challenged Southerners' image of themselves.[10]

The near complete absence of Latinos in these regions prior to the 1980s meant that for the lucky, ambitious, or clever few there were opportunities to make money servicing the needs of other immigrants. There is, in this sense, an Amador and a Los Tres Hermanos in every southern town with a Latino population. Were Amador's story to end here, he would simply have been an example of a successful immigrant: from poor Mexican farmer to Florida field hand to poultry plant worker to small businessman, a truly inspiring story. Amador, however, did not just sell chile ancho, corn flour, piñatas, compact discs, and other goods and services at Los Tres Hermanos. He sold Mexicans and, increasingly, Guatemalans. He trafficked in illegal immigrants while selling them fake Social Security cards and other phony documents. By itself, this would make Amador exceptional, though hardly unique. If a town's Latino population is large enough to support a Los Tres Hermanos, then it is almost certain that this sort of store will either provide these kinds of services or know where to arrange for them. And what makes Amador's case interesting is that he got caught, and caught in a big way.

In the mid-1990s, two Shelbyville police officers, William Logue and Donald Barber, began to question the immigration and Social Security cards that Latin Americans displayed when pulled over for routine traffic stops. When asked where they got the documents, many immigrants responded, "Los Tres Hermanos." The two officers, suspicious, contacted the INS. In 1997, an undercover INS agent purchased counterfeit cards directly from Amador and subsequently arranged with the unsuspecting grocer to bring undocumented workers to Tennessee and set them up with phony papers.[11]

The case gets more interesting. According to Kevin Sack, a journalist writing for the *New York Times*, Amador mentioned to the undercover agent that his former employer, Tyson Foods, regularly asked him to supply illegal immigrants. This led to a 1998 meeting at the Shelbyville plant in which Tyson managers asked Amador and the undercover agent to supply the company with two thousand illegal Guatemalans. For each

worker, Amador and the agent were paid (with corporate
checks) $100 or $200 for "recruitment fees." It is unclear how
many immigrants were actually supplied, or how much money
Amador earned off the smuggling venture, but by the year
2000 he had amassed a small fortune, including two stores, five
houses, and a number of cars.[12]

The downfall of Amador's little empire led to the largest
immigrant-smuggling case ever against an American corporation. In December 2001, the two-and-a-half-year investigation
came to a close. Led by the INS, and with the participation of
the Department of Agriculture, Social Security Administration, U.S. Attorney's Office for the Eastern District of Tennessee, the FBI, and several local police agencies, the prosecu-
tion convinced a federal grand jury in Tennessee to hand down
a thirty-six-count indictment against Tyson Foods Inc., two of
the company's executives, and four of its former managers.[13]

The Justice Department claimed that fifteen Tyson Foods
plants in nine states had conspired since 1994 to smuggle illegal immigrants across the Mexican border and set them up
with counterfeit papers. According to the indictment, the hiring of illegal alien workers was condoned at the highest levels
of management to meet production goals, cut costs, and maxi-
mize profits. Undocumented workers were preferred because
their fear of deportation meant they would accept working
conditions that U.S. employees would not. They would toler-
ate faster line speeds, take fewer bathroom breaks, and complain less to managers or government officials. They were also
less likely to file workers' compensation claims or be absent
from work. Fear of deportation produced the ideal worker.
This sentiment is hardly unique to poultry. A supervisor at a
meatpacking plant explained: "I don't want them after they've
been here for a year and know how to get around. I want them
right off the bus."[14]

The case against Tyson has its ironies. With great predictability, the federal government embraced its high-profile
role as corporate watchdog. Assistant Attorney General Michael
Chertoff reminded everyone that the "Department of Justice is
committed to vigorously investigating and prosecuting companies or individuals who exploit immigrants and violate our
nation's immigration laws. The bottom line on the corporate
balance sheet is no excuse for criminal conduct."[15] INS commissioner James Zigler, not to be outdone, noted that "companies, regardless of size, are on notice that INS is committed
to enforcing compliance with immigration laws and protecting America's work force."[16] In reality, neither agency has done
much to protect low-wage workers of any nationality, and both
have systematically ignored the issue of undocumented workers in certain low-wage industries.

Tyson's response was equally predictable. Ken Kimbro,
Tyson's senior vice president of human resources, was quick to
deny the government's charges: "The prosecutor's claim in this
indictment of a corporate conspiracy is absolutely false. In reality, the specific charges are limited to a few managers who
were acting outside of company policy at five of our 57 poultry
processing plants." Kimbro could hardly contain his indigna-
tion: "We treat all team members fairly and with dignity. We're
very proud of our diverse work force and encourage every
team member to express freely any concerns or questions they
have. We find it offensive that the prosecutor suggests that we
have treated any team member in a 'less humane' or a discrimi-
natory fashion. This is simply not true."[17] In an industry whose
labor force is thought to be one-quarter, one-third, or even
one-half "illegal," it is amazing that Tyson's upper management
could deny any knowledge of undocumented workers within
its plants.[18]

Yet what else could the government or Tyson say? Illegal

immigration is, after all, illegal. Neither Tyson nor the federal government could acknowledge what everyone knew to be the truth—that poultry companies routinely hire undocumented workers and that the INS regularly looks the other way. At best, Tyson could agree that a few undocumented workers slip through the cracks, admit that some might have been recruited by a handful of rogue managers, or simply plead helpless: it is, after all, difficult for poultry companies to determine whether every Social Security card or document is in fact legal. Tyson could never state publicly that it needed undocumented workers to deliver cheap chickens to American consumers and remain competitive in the poultry industry.

Similarly, it is expecting too much of the federal government to admit that it looks the other way precisely because the industry and America's chicken supply are heavily dependent on undocumented workers (or, if you are more cynical, because agribusiness is so powerful that the INS is forced to ignore the issue). This is the central irony and contradiction of the "immigrant problem." Until recently, it was unpopular for political leaders to embrace immigration, yet many of their corporate contributors (and virtually all consumers) depend on immigrant labor. The only interesting questions—Why Tyson? and Why now?—were left unanswered.

The case ended predictably. Even with the confessions and testimony of three Tyson managers (one of whom committed suicide), the government could not establish that the highest levels of Tyson management were involved in or even aware of the smuggling conspiracy.[19] There was no smoking gun. According to Tyson, a few rogue managers had acted on their own and were fired immediately when the scam was uncovered. In the absence of a paper trail, the jury's hands were tied.

The "immigrant problem," so central to America's food

supply, may make fools of companies and federal agencies, but it has much more serious consequences for the immigrants themselves. Although this case is extreme, with one immigrant literally selling another, it nevertheless highlights the tensions and divisions that have developed in Shelbyville, Tennessee; Rogers, Arkansas; Wilkesboro, North Carolina; and other relatively nonmetropolitan areas of the South where Latinos have settled. In these places, sometimes collectively called the Nuevo South, the opportunity for long-term upward mobility is severely restricted by limited labor markets. The chicken industry is often *the* option for a racially marginalized labor force with limited English skills—unless such job seekers manage to "rise up" into the business of servicing, and sometimes exploiting, fellow immigrants.

Many new arrivals, especially those who lack papers or who have relatively few connections, find it difficult to get settled and achieve any sort of working-class stability. These people tend to cycle in and out of poultry plants, often working in the worst jobs and worst plants, remaining marginal to even their own immigrant communities. In the case of the Tyson smuggling case, this group of marginal workers had a particularly ironic fate. According to Matthew Baez, who provides social services for immigrants in Tennessee, dozens of workers who were recruited in the Tyson scam "had no clue that was a scam or illegal. They were told these documents were for employment at Tyson and not to show them anywhere else."[20]

Most Latinos in the small-town poultry South fall somewhere in between the Amador entrepreneur and the most marginal of workers. Poultry workers are decidedly working-class, but in the past decade many have achieved relative stability, are permanently settled, and have acquired the trappings of middle-class life by working extremely hard, living frugally, and pooling multiple working-class incomes to support a single household. Unlike Amador, most in this group have been unable to become entrepreneurs and instead remain stuck in poultry, an occupation that makes it difficult to reproduce their way of life over the long term. Members of this group have transformed the South by making it their home.

A New Kind of Immigrant

One of the most interesting things about the Tyson case was how dated it seemed by the time the indictment came down in 2001. By the year 2000, few poultry companies were recruiting Latin Americans on a large scale. Most Latin Americans who found their way to the industry had prior work experience in the United States and were more likely to be coming from California or Texas than Mexico or Guatemala. Those workers who came directly from Latin America were generally recruited by family members rather than by poultry companies or their agents.

In fact, few Latinos arriving after the year 2000 fit the popular image found in media accounts of the naïve immigrant who is easily duped by the slick recruiter. Most have their own perspective on the process.[21] Indeed, Arturo, who came early and actually was approached by a recruiter, was savvy about the system:

> I was one of the first Mexicans in this area [of North Carolina]. I was recruited by a contractor who worked for [a poultry company]. He told me I would be able to save all this money—that the wages were high and housing cheap. And that the work was easy. I knew it was not true. I had already worked in Cali-

fornia, Texas, and Washington [State]. But it was either come here or go back to California. And California wasn't working. . . . [*laughs*] There are too
many damn Mexicans in California. I'm not kid-
ding. We are all trying to get the same jobs.[22]

Yet even if the workers know the game, they have few alternatives. A free bus ride and immediate employment means
they can hope to get ahead. As Arturo explained: "When I [ar-
rived in the United States] I [had] little money. The trip is very
costly. So I need to get a job immediately. I cannot spend two
weeks in a hotel waiting for the right job. I take what I can get.
Later, after I am settled, I can try to get something better."[23]

The early pioneers, mostly men, are lured even more by
rumors than recruiters. Subsequent immigrants arrive on the
more reliable advice of friends and family. Gustavo, for example, is

the Christopher Columbus of Arkansas. That's what
they call me because I discovered Arkansas for my
pueblo [in Mexico]. I came to Arkansas in 1987 or
1988. This was before anyone had come here. I had
been working in California for more than twenty
years. Each year I would come to California for
seven or eight months a year. At first it was just myself. Then I brought my sons. But never my wife. It
was no place for a family. We would live in a garage
with twelve men, each paying $100 a week. Can you
imagine? That's not for women and children. Every
year we did this, crossing the border, getting sent
back, crossing again, then picking fruit, working
in lettuce. . . . And every year we would return to

Mexico with enough money to live on for a few months before returning. We were surviving, but nothing else.

So then this friend, a fellow worker, said he had heard there were jobs in Arkansas in poultry. [*laughs*] I said, "Where's Arkansas?" and "What do you mean working in chicken?" I thought I would be working on a farm. We [immigrants] hear these rumors all the time. But then someone else said the same thing. So I talked to my sons, got some money, and went with one of my sons while the other two kept working in California.

We drove straight through. We were in a hotel the first week. There was no one to stay with during this time. The first day Tyson hired us. There were ten Mexicans in the plant. Pure gringos. They would play rock music while you worked. Now it's all Mexican. But they don't allow music anymore.

Right away I told my two other sons to come to Arkansas. I said quit your job [in California] tomorrow and come. They were here at the end of the week. The first two years we rented a trailer. Then my sons and I brought our wives and kids. Within four years we bought a house. Then the whole town [in Mexico] stopped going to California and started coming to Arkansas.[24]

Pioneers such as Gustavo pave the way for subsequent immigrants, many of whom follow their fathers, brothers, and friends to the promised land of Arkansas, North Carolina, and Georgia. Higher wages, year-round indoor employment, forty-hour workweeks, and relative job security make the decision to leave seasonal migrant labor an easy one. Hundreds of thousands of Latinos left California during the 1980s and 1990s. The clincher, as Gustavo suggests, is the lower cost of housing. Reasonable rents and affordable home prices make family living possible in places like Arkansas and Georgia in a way that it never had been in California. Lucia, a Mexican woman nearing sixty, put it this way:

> For years the men went to California. Once or twice they went to Texas, but usually California. I never went. I stayed with the kids, caring for the house [in Mexico]. I could have done the work. I can work in the fields. But it was the living conditions. My husband would not let me come because he said they lived like animals. That's what he said. Six or seven men in a room. Once my son's wife went. She said it was horrible.
>
> So one year he returns [to Mexico], like always. He comes home and says he went to Arkansas. I thought Arkansas was a city in California! He said they had a trailer and I was to come back with him. He said to start packing; we were leaving in a week! I had always wanted to come and now I was scared. . . . My daughters came, my sons' families came. Not all at once. But bit by bit. And we all got jobs in poultry. And little by little everyone bought their house. I know now how much the men struggled in California. Here [in Arkansas] we struggle as well. But we also advance. That is the difference.[25]

Both in conception and practice, the switch to poultry and to working in the southern United States is a story about upward mobility. Many of these pioneers have harrowing accounts of crossing the border, almost dying, and getting caught by border agents (*la migra*). But these stories are often seen as part of the past, recounted like old war stories. As the upwardly mobile immigrant becomes legal, the border loses some of its mystique and Mexico's attraction begins to wane.

> I crossed everywhere along the border. San Diego, Arizona, Texas, everywhere. In the 1980s I got caught crossing near San Diego three times before I got in. So then I tried Arizona. That was a terrible idea! We went twenty-three days with almost no food and little water. And no bath! We almost died, lost in the desert. We were robbed, beaten up. It was horrible. Now I have papers so I come and go like I want. But you know, now I don't even really want to return.

For the undocumented, the border works to keep them "in."

> It's hard to go to Mexico without papers. When I worked in California I had to go home every year. We didn't have a place to live or stable employment. So every year we had to face the border. Sometimes we were lucky, other times not. I was lucky. I never got raped or anything. But I did get robbed by a coyote, and *la migra* caught me three times. . . .
>
> Since we moved here [ten years ago], I have gone back once. It is not worth it. Here, I have papers to work, but I cannot leave the country legally. So I stay. I have a job, a house. My kids go to school here. No one bothers me. I don't know, maybe the

INS will pick me up one day, but I don't think about
it. I have almost ten years [in poultry]. Really, there
is nothing in Mexico for me. But still, I would like
to return.[26]

Tighter border control, which began in the mid-1980s,
has done little to stop illegal immigration. Between 1986, when
the Immigration Reform and Control Act was enacted, and
1998, INS funding increased eightfold and Border Patrol funding sixfold. Yet, the number of illegal immigrants doubled during the same period and is growing by an estimated 250,000 a
year. What tighter border control did do was make the passage
more expensive and more dangerous for immigrants. As a result, once immigrants finally arrive in the United States they
stay longer. Prior to the shift, immigrants stayed an average of
two and a half years, often cycling in and out of the United
States. By the year 2000, the average stay was over six years and
growing longer.[27] The circular pattern of temporary migration
has given way to permanent and semipermanent settlement.

Nevertheless, the upward mobility that poultry facilitates
comes at a tremendous price, even for those who have achieved
some level of working-class stability. Immigrants' relationship
to their Mexican communities becomes increasingly difficult
to maintain. They live as a foreign, dark-skinned, "other" that
is isolated and at times targeted by mainstream America. And,
above all, processing chicken destroys their health.

Esteban, from Veracruz, Mexico, took a job in "live hanging" at
a poultry plant in Bay Springs, Mississippi. If a company town
still exists in the twenty-first century, Bay Springs is it. Peco
Foods employs about eight hundred of the thousand residents
of this small southern town.[28]

Esteban's job in live hanging pays slightly better than other jobs, but it is a tough way to earn a living. As I described in the Preface, in one poultry plant where I worked, one of the largest slaughterhouses in the world, the live hanging area was massive. In an almost completely dark room, a handful of workers wrestle live chickens onto metal hooks that transport the birds through the plant. The workers move with incredible precision. Each of their arms operates independently and swiftly to pick up flailing chickens. In what appears to be complete chaos, the workers are supremely calm. Their motions are so rehearsed that each worker is able to grab two frantic chickens (one in each hand), hang them on the line, smoke a cigarette (without their hands), and heckle the new recruits as they watch in amazement. As the guy next to me said, "This is the most awesome thing I have ever seen." Few people last very long in live hanging, which is probably a good thing. The fingers and knuckles of one eight-year-veteran I talked to were swollen to the point of being almost unrecognizable.

Esteban was the archetype of a live hanger: young, lacking proper documents or English skills, hungry for success, and ready to earn cash. After he had been on the job for a year, however, his manager informed him that Peco Foods had recently received a letter from the Social Security Administration alleging that Esteban was using another worker's Social Security number. At first Esteban thought he would be fired and deported. As he conveyed to freelance journalist Russell Cobb, the manager "told me he could have the cops here in five minutes if I didn't cooperate with him."[29]

As it turns out, the manager was not interested in losing Esteban, or even concerned about document fraud or illegal immigration. Instead, "Gordo" simply informed Esteban that he had to get a new ID and Social Security card. Esteban was nervous. He lacked the money for new documents, was wor-

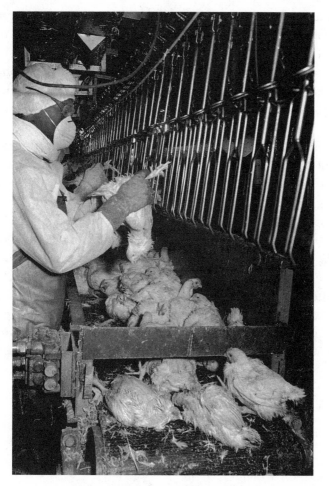

Live hanging (Photo by Earl Dotter)

ried about implicitly admitting that his old ones were fake, and didn't want to lose his seniority at the plant by starting over with a new identity.[30]

Gordo had a solution. As he had done for many other immigrants, Gordo offered to sell Esteban a new Social Security card for $700. He also wanted a date with Esteban's cousin. Un-

willing to pimp his cousin, Esteban contacted his union representative. Journalists discovered later that Esteban was not the
only one to find himself in this situation. Gordo would first
charge workers for their jobs, subsequently inform them about
the "no-match" letter from Social Security, and then request
additional payment for new documents. There were two discouraging ironies about the scam. First, Gordo was actually
the replacement for a manager whom the union had forced
out for similar abuses. The practice, it appears, was virtually
institutionalized at Peco Foods. Second, Social Security has no
law enforcement powers and does not share its data with
immigration agencies. The letters should not result in firing,
deportation, or extortion; companies are supposed to allow
the employee to handle the issue without interference. In practice, however, managers often fire workers for receiving such a
letter—sometimes with the implied support of local law enforcement, which has been known to threaten "illegals" with
deportation.[31] As a result, most immigrants in this no-win situation just keep their mouths shut, pay up, and struggle along.

VI
Inside a Poultry Plant

I arrive at Tyson's northwest Arkansas job center in Springdale at ten in the morning. Springdale, located at the center of the most productive poultry-producing region in the world, is home to the corporate headquarters of Tyson Foods. The Tyson job center itself is a small, unimpressive building with a sparse interior that resembles a government office. Signs surround the secretary's desk. In Spanish one says, "Do not leave children unattended"; another warns: "Thank you for your interest in our company, Tyson Foods, but please bring your own interpreter."

The receptionist seems genuinely surprised by my presence. "Sorry, hon, there are no openings for a mechanic. Fill out the application and we'll call you." Somewhat puzzled, I tell her that I am hoping to get a job on the production line at one of Tyson's nearby processing plants. With a look of disbelief she hands me a thick packet of forms and asks, "*You* want to work on the line?"

As I turn to take a seat, I begin to understand her confusion. The secretary and I are the only Americans, the only white people, and the only English speakers in the room. Spanish is the language I hear most, but it is not the only one being spoken. Lao is coming from a couple in the corner, and a three-some from the Marshall Islands is speaking a Micronesian language. Indeed, the U.S. South has become a key site for immigrant workers. Latin Americans, attracted by employment opportunities in the poultry industry, first began to enter northwest Arkansas and much of America's heartland in the 1980s. Today, about three-quarters of plant labor forces are Latin-American, with Southeast Asians and Marshallese accounting for a large percentage of the remaining workers.[1]

Tyson processes job applicants like it processes poultry. The emphasis is on quantity, not quality. No one at the job center spends more than a minute looking at my application, and no single person takes the time to review the whole thing. There are few pleasantries, but there is also no bullshit. I am spared questions like: What are your career plans? Why do you want to work in poultry? How long do you plan on working here? Instead, efficiency rules. Bob begins my "interview" with, "What can I do for ya?" I tell him I want a job at a processing plant, he makes a quick call, and in less than five minutes, I have a job on the line. I pass both the drug test and the physical. I am Tyson material.[2]

I arrive at the plant the following Tuesday, ready for work. The plant is massive, and its exterior is much like the job center: nondescript and uninviting. At 3 p.m. sharp, Javier, my orientation leader, gathers up the new recruits and escorts us into a small classroom that contains a prominently displayed sign: "Democracies depend on the political participation of their citizens, but not in the workplace." Written in both English and Spanish, the message is clear in any language.

Of the nine others in my orientation class, eight are Latin Americans, with six coming from Mexico and two from El Salvador. As the men in the plant frequently lament, women workers tend to be slightly older. In this respect, the two women in our group—Maria, in her early forties, and Carmen, in her early fifties—are typical. The six men vary considerably in age. Juan, from El Salvador, is only twenty-three, but Don Pablo is well into his sixties. Jorge, in his mid-thirties, has spent the past thirteen years in California, working mostly in a textile factory. Although he has only been in Arkansas for a few days, he already appreciates the region for many of the same reasons that draw other Latin Americans. Like Jorge, most of the Mexican workers come from rural areas in the state of Guanajuato, pass through California where they work in factories or pick fruit, and then find their way to the promised land of Arkansas. Not only is everything in Arkansas much cheaper than in California, but Tyson Foods also pays around eight dollars an hour, offers insurance, and consistently provides forty hours of work a week.

After putting on our smocks, aprons, earplugs, hairnets and/or beard nets, and boots, we begin the tour of the plant. No one is unaccustomed to hard work, and most have killed chickens on farms in Latin America, but the combination of sounds, sights, and smells that we experience next is overwhelming. It does not help that the tour begins in "live hanging," where, as I described in Chapter 5, live chickens are hooked upside-down by their feet to an overhead rail system that transports the birds throughout the plant. As we watch, blood, feces, and feathers are flying everywhere. Carmen says what we all are thinking: "My God! How can one work here?" The answer, it turns out, is simple: Live hanging pays a bit more than the other plant jobs.

Fortunately, I land a job on the saw lines. It's not exactly pleasant, but it's a long way from live hanging. These further-processing lines are at the heart of the revolution that has transformed the poultry industry and American diets over the past twenty-five years. Close to my work area is the "Church's Line," and the next summer I will labor next to the "KFC Line." Where I am, the task is relatively simple. Each processing line takes a whole chicken, cuts it, marinates it, and then breads it. With twenty to twenty-five workers, each line processes about eighty birds a minute, or forty thousand pounds of chicken a day.

My coworkers are an interesting and diverse bunch. Of the twenty or so workers who keep the lines running, only three of us are white Americans. Most local white workers left the plants during the region's economic boom of the 1990s, and those who remain tend to fall into two categories.[3] An older group has been working at Tyson for more than twenty years. They have found a niche and hang on to the benefits that seniority bestows. Then there are the few white workers who started at Tyson more recently. They are usually there because poultry is one of their only options. Jane, for example, is well into her sixties and had worked at the plant before Tyson bought it. She subsequently moved out of the area and spent most of her life working in a factory that produced medical equipment. After her husband died, she returned to northwest Arkansas, walking to the plant every day from her small apartment. Factory work is all she knows.

Most of those who work directly on the line are women, often "older" women in their forties and fifties. Jane, along with Alma, Gabriela, and Blanca, usually sorts chicken. Alma and Gabriela are Mexican sisters in their forties. Blanca, also from Mexico, has a husband and four children working at Tyson, including Maria, who checks the marinade. Their counter-parts on the other line, Li and Lem, are both from Laos, in their fifties, and couldn't be more different. Lem is friendly, always willing to help out coworkers. Li has two personalities. On the plant floor, she barks out orders like a drill sergeant, seemingly unaware that no one can understand her. In the break room, she is gracious and gentle, sharing her culinary delights with Laotians, Mexicans, Salvadorans, and hungry anthropologists alike.

The fact that most "line" workers are women is neither coincidence nor insignificant in a plant where about two-thirds of the workers are male. Line jobs are the worst in the plant. They are not only monotonously repetitive; they are dangerously so. When the line is working properly, line workers can hang chickens at a pace of forty birds a minute for much of the day. They stand in the same place and make the same series of motions for an entire shift. If you stay in these jobs long enough, you will inevitably develop serious injuries associated with this repetitive motion.

By contrast, although auxiliary workers tend to do the same tasks all day long, they do not do the exact same series of movements with nearly the intensity as line workers. Women tend to be excluded from these auxiliary jobs because they often involve heavy lifting or the operation of machinery. Mario, Alejandro, Roberto, Juan, Jeff, Carlo, and I are "young" men who clean up waste, bring supplies, lift heavy objects, and operate handcarts and forklifts. As auxiliary workers, we do line work, but only intermittently.

I am to be the *harinero* (breading operator), or, as my twenty-two-year-old supervisor Michael likes to call me, the little flour boy. Michael's instructions are simple: "Do what Roberto does." But Roberto provides little formal training.

With five years on the job, everything is natural to him: he can do every job on the line, fix the machines, and carry on a conversation all at the same time. So I just watch, hoping to gain his respect and pick up anything that will allow me to survive the first week.

I learn quickly that "unskilled" labor requires an immense amount of skill. In one sense, the flour boy's job is straightforward: it entails emptying fifty-pound bags of flour all day long. The work is backbreaking, but it requires less physical dexterity than many of the jobs on the line. Yet the job is also multifaceted and cannot be learned in a single day. Controls on the breader and rebreader need to be continually checked and adjusted, the marinade needs to be monitored, the power needs to be turned on and shut off, and old flour needs to be replaced with fresh flour. All of this might be manageable if the lines functioned properly, but they never do.

The rebreader is the source of more than 75 percent of all the lines' problems, particularly those that force a shutdown. It is here, with Roberto, that my education as flour boy begins. One of the first things I learn is that I will be doing the job of two people. There have always been two flour boys, one for each line. But our supervisor, Michael, has recently decided to operate both lines with only one flour boy. He is essentially doing what he has done, or will do, with virtually all of the jobs on the saw line. Where there used to be three workers hanging chicken, there are now two. Where there used to be three or even four workers arranging parts, there are now two. Whereas two people used to check the level of marinade, now there is only one. Nor is this downsizing limited to our section of the plant. About six months prior to my arrival, an older generation of supervisors, most of whom had come up through the production lines, was essentially forced from their jobs when a

Inside a processing plant (Photo by Earl Dotter)

new set of plant managers took over. The new managers had made the older supervisors push the workers harder and harder. The supervisors, who knew what it was like to work on the line, eventually refused by simply leaving the plant. As a result, a group of younger, college-educated supervisors was brought in.

Michael is part of this younger generation of supervisors. One of the first in his family to attend college, Michael, when I met him, had just graduated from the University of Arkansas with a degree in poultry science. Although he had "never imagined" earning so much in his first year out of college (supervisors start at less than $30,000 a year), the trade-off was considerable. Michael had no life outside of the plant. He arrived every day at 12:30 in the afternoon and never left the plant before 3:30 in the morning. Unlike line and auxiliary workers, he enjoyed a job with some variety, almost never got his hands dirty, and could hope to move up the corporate ladder. At least in the short term, however, he was tied to the plant just like the

Worker hanging chickens (Photo by Earl Dotter)

rest of us. Nevertheless, it was Michael who was the focus of our anger and who (guided by his bosses) was implementing the latest round of downsizing.

Part of the reason why the downsizing was possible, besides the lack of a labor union or binding job descriptions, was that reducing the number of line workers does not necessarily prevent the line from running. The fewer workers simply have to work faster to keep pace with the line. As Roberto was quick to point out, the position of the breading operator is somewhat different. When the breading operator does not keep up, the entire line is brought to a standstill. And Michael was replacing two experienced flour boys, Roberto and Alejandro, with a single trainee—me. As Roberto explained: "When Michael told us he was going to only have one flour boy, we were not totally surprised. I told him I was quitting as flour boy and would work on the line. Alejandro left [as flour boy] in less than a week. Michael couldn't find anyone to take the job. . . . It was too much work. So he had to get a new guy who couldn't say no, someone like you."[4]

Roberto's understanding of the situation was absolutely correct, but he was being less than honest. It *did* matter to him. Alejandro was more forthcoming: "I had eight years as flour boy. I like the job. It's like family here. It doesn't mean anything to Michael. For him it's just a job and we're just Mexicans. He doesn't know anything anyway. I wanted to stay, but why? Twice as much work for the same salary. I did my job well. I have nothing to be ashamed of."[5]

During my first weeks on the job, the line is continually shut down for one reason or another. Few of the problems have anything to do with me, but the entire process is slowed by the fact that there is only one real flour boy—Roberto. The flour boy has to fix just about everything, but the central problem is that the rebreader simply does not have sufficient power to circulate the flour through the apparatus while pushing the chicken along the conveyor belt. In short, when one opens the valves and allows enough flour to flow through the machine in order to bread the chicken, the machine bogs down, the chicken piles up, and the parts begin to fall on the floor. This results in

loud shrieks from just about everyone on the line. The breading operator, in this case Roberto, has to then shut down the
entire line and figure out exactly which part of the rebreader is
malfunctioning.

The situation creates an ongoing struggle between Michael,
as supervisor, and Roberto, and me. Several solutions are possible. First, and most obviously, the mechanics could feed
more power to the machine, thus giving it the capacity to handle the weight of the chicken and flour. This is clearly what
Michael wants. Second, we could run less chicken, which, by
reducing the weight on the conveyor belts, would allow the rebreader to operate properly with the existing amount of power.
From Michael's perspective, this is simply unthinkable. Our
guiding principle is to keep the line running at all times, at
maximum speed, and at full capacity.

Consequently, Roberto and I adopt two strategies in order
to keep the rebreader working properly and the lines running
smoothly. First, we change the flour frequently. Fresh flour
that has not yet become moist and clumped together from the
wet chicken is lighter and circulates more smoothly through
the entire apparatus. Michael, however, does not like this option because it is more expensive. Second, we use only as much
flour as the rebreader can support. But here again Michael in-
sists both that the rebreader can handle more (old) flour and
that the levels we run it at are inadequate to bread the chicken
sufficiently.

If Michael's being wrong weren't frustrating enough, his
style of managing the problem is exasperating. He passes by
every hour, sees there is not enough flour on the conveyor belt,
and tells Roberto and me to increase the flow. Confident that
he has set us straight, he then leaves, and with uncanny timing
the machine soon bogs down. Roberto and I have to stop the

production line, clean the mess, and readjust the flow of the flour so the machine can run without stopping. Michael then returns, wonders why there is not more flour on the conveyor belt, and the whole cycle begins again.

This uneasy and somewhat absurd tension characterizes the workday from beginning to end. Only occasionally would Michael be present when the rebreader bogged down as a result of his own miscalculations. Roberto and I relished these times. Roberto would suddenly "forget" how to fix the machine, so he could enjoy watching Michael frantically call a mechanic for help. The mechanic would eventually arrive, talk to Michael, stare at the machine for several minutes, and then swallow his pride and ask Roberto what the problem was. Roberto would then look at Michael, smile at me, and fix it in a matter of seconds.

Looking back, I find it hard to explain why this petty struggle seemed so important at the time. Of course, it would have been in our interest to just follow Michael's misguided directions and let the rebreader bog down and the production line stop. It was a pain to continually fix the machine, but we got paid the same amount regardless of whether we were emptying bags of flour, fixing the rebreader, or standing around talking. Moreover, a shutdown generally proved that Michael was wrong, something that Roberto and I took great satisfaction in even as we were simultaneously annoyed that the lines weren't running. Finally, every other worker in the group was able to have a break when the saw lines were shut down.

Why, then, did Roberto and I, as well as the line workers, become profoundly irritated when one or both of the saw lines shut down? For one thing, the downsizing had angered everyone and confirmed our collective perception: Michael's lack of experience led to decisions that made our lives intolerable and that were economically unsound. Second, and perhaps most important, Michael was degrading the work of the flour boy by ordering Roberto and me around. He was keeping us from controlling our own labor, a control that had given the job meaning. Finally, virtually all of the workers took great pride in their jobs. To be sure, there were times when we enjoyed a shutdown. The more frantic Michael got, the more we celebrated. Most of the time, however, we were frustrated by having a "bad" day, even as we relished the agony that a stoppage would cause him.

Despite our protests, Michael forges ahead with his plan, and on the Monday of my fourth week I begin running both lines myself. What he does not tell us, however, is that he finally has enlisted the mechanics to boost the power going into the production lines. When Roberto and I arrive that Monday, we know Michael has won. With increased power, the rebreader runs smoothly and almost never bogs down. For Roberto, this means that running the lines no longer requires his expertise. Now the lines run not only with fewer problems, but faster and with the capacity to handle more chicken. For me, it means that the job involves less skill, but much more work. I fill the flour not just for two lines, but for two lines running more consistently and at a faster pace. With the sometimes entertaining struggle now resolved, the intensity and monotony of the job become almost unbearable to me. As for the rest of the workers, particularly those on line, the change is devastating. The lines shut down less, there are fewer breaks, and the pace is quicker. By the end of the week, Blanca, a Mexican woman in her fifties, is overwhelmed. She has been hanging chickens for too many years, and her body simply cannot withstand the increase in work. Although she had hoped to stay at Tyson until retirement, she is forced to quit within the week.

A Break

Communication and self-expression among workers are muted on the plant floor by the intensity of the work, the noise, and the supervision. Knowing glances, practical jokes, cooperation, and shared pain become ways that workers acknowledge, in quiet ways, their shared experience. In the cafeteria or break room, however, people feel freer and more relaxed. Twice an eight-hour shift, for thirty minutes, workers watch Spanish-language TV, eat and exchange food, complain about supervisors, and unwind. More often than not I was the only American in the main break room. The few American workers on the second shift were almost always congregated in a smaller break room where smoking is permitted and the TV is in English. Supervisors also almost never enter the break room where mostly Spanish is spoken, and when they do they are noticeably uncomfortable. At least here, then, the inmates are in charge.

On this particular day, Michael had pushed us hard and had brought free boxes of fried chicken in order to thank us. It was a gesture that he would do half a dozen more times while I was working in the plant, and it always produced roughly the same reaction from the workers. We would look at the chicken, stare at each other, and someone would say, in Spanish, something like: "Asshole. I am not going to eat this shit." Next there would be an awkward moment when we would quickly look at each other, look away, and pretend not to know what was going on. Then someone would say: "We can't throw away good food, and we're all hungry. Let's eat this shit." And we would angrily grab the chicken, eating most of it or carefully packing it away to take home to family.

Michael's gesture was insulting for many reasons. First, he wasn't just giving us food; he was giving us chicken. Second, the gift didn't come close to making up for what the workers had just experienced on the plant floor. Third, it was pathetically and transparently paternalistic. And finally, Michael's token embarrassed us because we knew, even as we hesitated, that we were not in a position to throw away food. We would process our chicken and eat it too.

As we sat eating the chicken together that day, no one ever directly talked about Michael's gesture, or our acceptance of it. There seemed to be a collective agreement not to relive the humiliation:

ROBERTO [*welcoming me into the group*]: "Ai, Steve, you are almost Mexican. All you need is a Mexican wife to cook you some decent lunch and you would be Mexican."

ALEJANDRO: "Yes, Steve is a Mexican. He speaks Spanish, eats with Mexicans, and he works like a Mexican. It's pure Mexicans here. We all eat chicken."

ELISA [*protesting kindly*]: "Ai . . . I'm not Mexican. I'm Salvadoran."

ALEJANDRO: "Look, we're all Mexicans here [in the plant]. Screwed-over Mexicans. [He *points to the Laotian Li.*] Look, even she is a Mexican. Pure."

[*We laugh as Li, who can't hear us, quietly eats a chicken wing.*]

ELISA: "Yes, it's the truth. We are Mexicans here in the plant, especially inside [on the floor] when we are working."

ME: "And outside the plant, in Fayetteville, Springdale, and Rogers? Are we all Mexicans outside?"

ROBERTO: "Outside, we are all fucked. We're in Arkansas."

[*Everyone laughs.*]

ALEJANDRO [*talking to me*]: "Outside, you're a gringo. You are from here. Outside, we are Mexicans, but it is different. We're still screwed, but in a different way. We are foreigners. We don't belong. At least here in the plant we belong even if we are exploited. Outside, we live better than in Mexico, but we do not belong, we are not from here and keep to ourselves."

ME: "And in Mexico? Who are we in Mexico?"

ROBERTO: "In Mexico, you are a gringo. You are a foreigner, but not like we are here in Arkansas. You are more like a tourist; treated well. We are not tourists here. We are treated more like outsiders. In Mexico, we are normal people, just like everyone else. Because it's all Mexicans. But in Mexico there is no future. My children were all born here, they are Americans. They have a future. Now, when I return to Mexico I feel like a tourist. I have money, travel, visit people. Our future is here now."

ALEJANDRO: "At least in Mexico the chicken has some fucking taste."

When Alejandro looks around a cafeteria filled with people from Mexico, El Salvador, Honduras, Vietnam, Laos, and the Marshall Islands and says that we are all Mexicans, he is making a statement about our shared experience as workers who do the lowest levels of work. In this respect, Li, who comes from Laos, is Mexican, one of us, because she does the same crap job as everyone in the room; because she eats Michael's chicken; and because she is Mexican to Tyson's all-white management.

This conversation suggests that we should at least consider the possibility that transnational migration may make people question the very categories that borders support. The experiences workers have in their new countries may encourage both immigrants and the native born to develop class-based notions of affiliation, identity, and loyalty. We must consider that globalization can lead not only to the internationalization of capital, but to the internationalization of workers in gathering places such as poultry plants. This does not mean that such sites will be similarly experienced or understood, or that they will automatically unite this diverse working class any more than factories did in nineteenth-century England. What it does mean is that if we are going to understand "transnationalism" in a more profound way, then we need to see culture not just in terms of ethnic-national rituals and customs but also from the perspective of class formation. The Mexicans, Salvadorans, Vietnamese, and Americans with whom I worked experienced cultural difference every day—when they befriended, cooperated with, and shared native foods with each other. But they also ate together chicken that was as painful to swallow as it was to process.

The conversation we had in the break room also suggests how important life outside the factory is for immigrants, and how difficult it is for work-based alliances and allegiances to stay intact after work hours. It is ironic, given the difficulty of the work, that many immigrants feel that the workplace is more nurturing than their local communities. Indeed, for most male workers, most of whom previously worked in Mexican and Californian agriculture, poultry processing is understood through the lens of upward mobility. For them, the difficulty of the work, as well as the low pay, are taken for granted and compare favorably with past forms of employment. Women workers, for their part, are often excited about having their own income, often for the first time, and the independence that it potentially brings. Moreover, for both sexes, social and even familial relations are often forged in poultry plants—relations that make life in the community both possible and meaningful. It is not uncommon for a conversation about work to turn into a discussion about how so-and-so met their husband or wife while processing chicken on the line. In this sense, many immigrant workers feel much more at "home" on the processing plant floor—working one of the less desirable jobs in America—than they do in communities where they live as a racially marked and conspicuously foreign working class.

Watching the Chickens Pass By

There is no shortage of eye-catching and often horrific stories when it comes to the meat industries. The deaths of some two dozen workers in a 1991 fire at an Imperial Food processing plant is perhaps the most tragic case, but it is by no means isolated. In my two summers at Tyson, however, there were no horror stories of the type that make the evening news. To be sure, safety regulations and standards designed to protect both workers and consumers were routinely broken or ignored in the poultry plants where I worked. But what impressed me most about the strange world of poultry processing was the unbearable weight of routine. The oppressiveness of routine work is very difficult to convey. Yet it defines factory life and is perhaps the most devastating part of work in the poultry industry. Minute by minute, hour by hour, day by day, month by month, year by year, one of the most basic features of life—work—becomes an unbearable and unwinnable struggle against the clock. In fact, the more one struggles, the worse it is. The first complaint that virtually all workers point to—before wages, working conditions, and supervisors—is the intolerable monotony:

> It's something that is impossible to describe. You worked here, so you understand. It's weird, but for

three or four days a week, at some point during the day, I honestly feel like I will not make it through the day . . . that I will not possibly make it to break. Most of the time I think about something else, or play a game with myself. I try to make the best cuts, or see how fast I can work, or see how little I can do and still get the job done, or see if I can do it with one hand—something, anything. But at some point during the day these little tricks don't work and I feel like I am going to have a panic attack. I look at the clock. Then I start to think about every movement I make. Once you start to think you are finished. It's like if you think about breathing, you can't breathe. I feel like I am going to scream. The clock does not move. I swear it goes backward! You know what I am saying. I even tell God that if he lets me make it to break I will never come back. I promise myself I will quit. And I am totally serious.

I have been playing this game for ten years! I come back every day. For ten years I have been torturing myself, spending the best years of my life in this ugly building, without windows, watching the chickens pass by, doing the same exact thing. I honestly don't know why I do it. The money of course. And once I leave the plant I somehow forget how awful it was and here I am the next day. I can't explain it. I suppose I am so relieved when I leave the plant that I forget how bad it was.[6]

Routine does not simply mess with your mind; it destroys your body. Not all workers are affected in the same way or to the same extent, but if you spend more than a year on the line—doing the exact same series of motions over and over again—it is certain that your fingers, wrists, hands, arms, shoulders, or back will feel the effects. A few more years and the damage may be irreparable. Almost all of the line workers I met had serious wrist problems. Many had undergone surgery and more than a few were permanently debilitated.[7] A 2005 Human Rights Watch report on the meat and poultry industries put it this way: "Nearly every worker interviewed for this report bore physical signs of a serious injury suffered from working in a meat or poultry plant. Automated lines carrying dead animals and their parts for disassembly move too fast for worker safety. Repeating thousands of cutting motions during each work shift puts enormous traumatic stress on workers' hands, wrists, arms, shoulders and backs. They often work in close quarters creating additional dangers for themselves and coworkers. They often receive little training and are not always given the safety equipment they need. They are often forced to work long overtime hours under pain of dismissal if they refuse. Meat and poultry industry employers set up the workplaces and practices that create these dangers, but they treat the resulting mayhem as a normal, natural part of the production process, not as what it is—repeated violations of international human rights standards."[8]

As one of my coworkers points out, the pain can cause even more anguish because "as soon as I start hanging chickens I feel fine. It's like that is all my muscles know how to do. I am in constant pain when I am not at work. My hands hurt so bad sometimes that I cannot make dinner or hold my child. When I wake up in the morning it takes my hands and arms thirty minutes to wake up."[9] In other cases, the work provides no relief: "This is the fifth or sixth job [in the plant] I have had. After a year in one job I can't do it anymore. Something starts

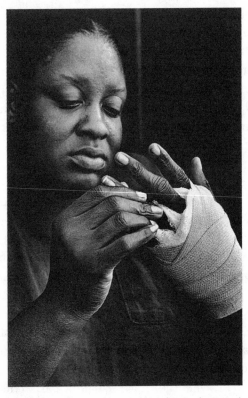

Injured poultry worker (Photo by Earl Dotter)

to hurt so bad that I can't do the job. If I complain enough they usually switch me, but not always to something better."[10]

The industry's response is that workers develop repetitive motion disorders and other injuries when they fail to follow proper procedures. If workers would just use tools properly, maintain good posture, and do their exercises, they would experience no pain or injuries. Such statements would be comical if the consequences were not so tragic. Further, the company's stance can be perverse. On one memorable occasion,

Michael, our supervisor, conducted a routine "training session" on ergonomics. Because the supervisors could not afford to stop the line, Michael was to read "the lesson" while the workers continued to work and the machines drowned out his voice. Each worker was to then sign a sheet of paper confirming that he or she had received the lesson. "I sign," one worker quipped, "because I do receive a lesson. I learn how little Tyson cares about us."

On this particular occasion, Arturo challenged Michael in full view of the other workers, suggesting that if the instruction was to have any meaning he had to stop the line and gather the workers in a quiet place for the lesson. Michael, embarrassed, reluctantly stopped the line, moved the workers into the hallway, and read a single sheet of paper with about ten points. Arturo insisted that I translate, since, as he pointed out, there was little point in conducting a lesson in a language that the majority of workers could not understand. (Even with the translation, four of the workers from southeast Asia were left completely out of the loop.) The ten points were all straightforward enough. Workers should use their legs, not backs, to lift heavy objects; they should stand close enough, and at the right height, when sorting chicken on the conveyor belt; and so on.

Hoping to avoid discussion, Michael ignored the last line of the lesson, "Ask the workers if they have any questions," which I blurted out in Spanish before he could send us back to work. The women line workers responded by approaching Michael with a ferociousness that caught everyone off guard. The scene must have looked a bit odd from a distance. Eight or nine Mexican and Salvadoran women, all over forty-five years of age and standing about five feet tall, were berating their bewildered six-foot-three, twenty-two-year-old supervisor. Looking at Michael with serious determination, Maria began, "Tell him we know how to do our jobs. But we are too short and need stools so we can be at the correct height." Ana chimed in: "Tell him I agree with the lesson. We shouldn't reach as far as we do. It hurts our backs. But if you are by yourself you have to reach [all the way across the conveyor belt]. The problem is we don't have enough workers at each station." Feeling momentarily empowered, Blanca quickly added: "I can barely move my wrists when I get home because I am doing the job of two people. We don't need a lesson, we need more workers. Tell him to come work and see what it is like."

I translated as quickly as possible. The women were serious, but they were also enjoying the moment. Things were how they should be. Michael was a kid receiving a tongue-lashing from women who were old enough to be his mother. Winking at me, Isabel said, "Tell the boy we know how to do our jobs and that he needs to start doing his." As we well knew, Michael *was* doing his job. That was the problem. Looking for an exit, Michael panicked and dug himself in even deeper: "Tell them if they hurt they should go to the nurse." There was hardly a sorer subject than the ineffective company nurse. All of the workers in the group started to laugh dismissively. To make sure he got the message, Maria scoffed, "When we go to the nurse she just gives us Advil and tells us to go back to work." She lifts her arms. "Look at my wrists. Do you think Advil is the answer?" And with that, the workers decided that the discussion was over and returned to work.

The production line and factory rhythm give work an unbearable routine that places mental and physical strain on workers of all ages, genders, and nationalities. There is no escaping it. Unfortunately, the weight of routine does not end at work. When I first began at Tyson I had this naïve idea that the one virtue of working in a factory was that once I left work I would be free. Little did I know.

I worked the second shift, arriving around 2:30 in the after-noon to set up the production lines. I would then lift bags of flour for the next eight or so hours, enjoy two half-hour breaks, prepare the lines for the cleaning crew, and then finally get out of the plant sometime after 12:30 a.m. (and often after 1 a.m.). Exhausted, I had to first shower and then wind down by hav-ing a beer and watching some late-night TV before hitting the sack. Rarely did I get to bed before 2:30, and I often fell asleep in front of the TV. I slept well but frequently woke up in the middle of the night with the sensation that my hands were so bloated they were going to explode. This is what you get when you clench bags of flour all day long. Around nine or ten in the morning I would wake up in time to do it all over again. Free time? I had trouble finding the time, let alone energy, to shop, exercise, or even do something as simple as get a haircut.

Yet my situation was much easier than that of the other workers. My stint in the factory was temporary, a fact that was not only comforting, but allowed me to postpone or ignore "life" in ways that most workers could not. I had no family, few commitments, and no financial problems. I came home to a quiet apartment and had the luxury of vegging out. No other worker could do that. Most had families and many worked an-other job. Some even additionally worked the night shift in our plant. As one worker recounted:

> I am always tired. The worst part is that since I work
> the second shift I am rarely around my children or
> wife. When I am home I am usually asleep. I get
> home at one in the morning. My wife works the

first shift so when I return she is asleep. We cannot both work the same shift because someone has to be with the kids. When I wake up in the morning she is already at work and I have to get the kids ready for school. Something always goes wrong. One can't find his shoe. Another lost this or that. It's a circus. I get them to school, come home, sleep a little bit more before I get ready for work. I go to work and my wife is leaving. She gets home in time to be with the kids in the afternoon. Sometimes there is a little time when the kids are alone. We don't like that because something could happen, but it is no more than an hour and the oldest is now thirteen. I see my wife on Sunday. We joke that it is a good thing we don't want more kids because we don't see each other enough to make them![11]

The sad irony is that although working in a factory imposes a routine on daily life, it is not one that makes it possible to have a "normal" life. Even the mundane tasks of daily living—shopping for groceries, putting children to bed, intimate moments with a spouse—become difficult to accomplish. It is hard to tell which is more overwhelming: the oppressive routine at work or the inability to establish a viable routine beyond the factory gate. For immigrants, as we will see in Chapter 7, the predictability of work is a source of pain, but it is also a refuge from the unpredictable world outside the factory doors.

VII
Growing Pains

Thousands of Mexican and Central American immigrants have moved to small towns in the rural United States during the past two decades. This fundamentally diverse population, loosely called "Hispanic" or "Latino," has rapidly transformed American society. Nearly half of the Hispanic immigrants living in the United States arrived during the 1990s alone. The Hispanic population increased almost 60 percent during this period, from roughly 22 million in 1990 to over 35 million in the year 2000, making this group the largest minority in the United States. (This figure does not include the estimated three million Mexicans who live in the United States illegally.) By the year 2050, it is estimated that one out of every four Americans will be Hispanic.[1]

California used to be the most common destination for Hispanic immigrants, but its saturated labor markets, outrageously high cost of living, and xenophobic legislation such as Proposition 187 have all made that state less attractive to immigrants. Over half of the nation's Hispanics still live in California and Texas, but almost one million left California for other states during the 1990s. Twenty-two states experienced a doubling (or more) of their Hispanic populations during the same decade, with southern states experiencing the fastest rates of growth.[2]

Many immigrants settled in urban centers of Middle America such as Atlanta, Memphis, or Iowa City, but thousands of others made their way to small towns throughout the South and Midwest. Although the recent flow of migrants began in the 1980s, relatively few small towns and rural counties in the heart of America had statistically significant Latin-American populations prior to 1990. By 2000, few had been left untouched. Populations swelled as the percentage of Hispanics reached one-quarter, one-third, or close to one-half of a town's total residents. In many places, this change took place within a decade; in some cases, it occurred within five years. Cabarrus County, North Carolina, for example, experienced an increase of 1,300 percent, from nearly 500 Hispanics in 1990 to over 6,500 in 2000. During the same period in Galax, Virginia, the number of Latinos went from 65 to 757, an increase of over 1,000 percent. Ligonier, Indiana, saw its number increase from 321 to 1,452 in a span of a few years.[3] And in Collinsville, Alabama, a small town of 1,500, Latinos started arriving in the early 1990s to work in a local poultry plant. By the end of the decade they numbered over 400.

This wave of immigration caught small-town America by surprise. Not wanting to attract attention, immigrants moved into segregated workplaces and neighborhoods where other immigrants had established a presence. Whites and blacks, when they noticed immigrants at all, figured "the Mexicans" were passing through. They were migrants. How and why would a group of non-English-speaking, brown-skinned foreigners settle permanently into towns and counties whose racial makeups seemed set in stone? It was hard to imagine.[4]

Work is clearly a fundamental part of the immigrant story. If there were decent-paying jobs in Mexico, few Mexicans would be living in Georgia or North Carolina. Hispanic men have the highest rates of workforce participation of any group in the United States. Even with high rates of poverty, Hispanics are on welfare less than other poor Americans, increasing their household incomes even though they tend to work in low-wage industries.[5] Hispanics, regardless of whether they are born in the United States or Mexico, or whether they are legal or illegal, work and work hard.

Throughout much of Middle America, the influx of Latinos coincided with an economic boom that left many of the lowest-wage and least pleasant industries searching for workers. Companies in the heartland needed workers, put the word out, and immigrants appeared. In some places, this demand for labor was increased by a general loss in local population. In other places, and in certain industries, the "need" for workers was largely a fiction created by companies that wanted a labor force that could be paid less, would work harder, and was more easily intimidated.

Latin Americans may come for work, but they stay for a range of reasons. The low cost of living in rural America is key, but so too are less tangible factors. More than any other part of the United States, the rural South resembles the communities that many Latin Americans left behind: these places are rural, family-centered, and religious, and have a slower pace associated with country life. These are the reasons that many immigrants give when explaining why they have made places like Siler City, North Carolina, and Rogers, Arkansas, home. The American South is, so to speak, a touch Latin-American.

The welcome that Latin-American immigrants receive upon arriving in Middle America varies considerably depending on time and place.[6] Race relations, economic opportunities, and other preexisting conditions shape how a particular town or region will receive immigrants. Even in the same town opinions vary considerably. For some, immigrants represent a much-needed labor force that will do the work that no one else wants. A few have even welcomed them as a source of cultural diversity, energy, and hope for small towns that have been in economic and demographic decline. Entrepreneurs have slowly come to realize that these working-class families represent an important new source of business. In North Carolina alone, Hispanic buying power increased from $843 million in 1990 to $8.3 billion in 2003.[7] Even many politicians, including Republicans such as President Bush, are actively courting the "Hispanic vote"—that is, when they are not banging the anti-immigrant drum. Others, however, have been less embracing. Immigrants have been blamed for just about everything, from lowering wages, increasing crime, and bringing down property values to destroying "American" culture and siphoning off public resources from schools, police forces, and social services. Immigrant bashing can be popular, especially on the local level around election time.

From Aunt Bee to David Duke

It was not all that long ago that Siler City, located at the geographic heart of North Carolina, was best known as the shopping destination for citizens of Mayberry, that small-town paradise immortalized on *The Andy Griffith Show*. In fact, when Frances Bavier, the actress who played the show's Aunt Bee, retired in 1972, she moved to Siler City because it reminded her so much of her fictional hometown. Siler City was the archetypical southern town. With close to five thousand people, it was neither too large nor too small. Located just far enough from Raleigh and Greensboro, Siler City provided access to urban niceties while retaining its small-town feel. A healthy mix of industry and agriculture kept the economy afloat and the downtown vibrant. And as late as 1989, Siler City's population looked like that of so many other small southern towns: black and white with no shades of brown.[8]

Change was on the horizon. Beginning in the mid-1990s, Latin-American men such as Wilfredo Hernandez began to arrive, find jobs in poultry and other local industries, and bring their families to settle in Siler City. Hernandez, who fled civil war in El Salvador, arrived in California in 1981. Sharing a house with thirteen roommates, Hernandez washed dishes from 5 P.M. to 3 A.M. for $3.35 an hour. In his precious spare time, he walked forty-five minutes to attend English classes. Wilfredo eventually married, but even with the additional income of his wife, Blanca, they could barely afford a one-bedroom apartment. They also feared urban violence and worried that their Los Angeles neighborhood was too dangerous for their growing daughter. As he conveyed to journalist Barry Yeoman, "California might have been the promised land twenty or thirty years ago. Not anymore."[9]

The American Dream, or at least a "real" job and a safe community, was tantalizingly close, however. Members of Wilfredo's family, including his mother, had recently moved to Siler City and were doing well. Wilfredo and Blanca were contemplating such a move when the California earthquake of 1994 left them and thirty thousand others homeless. That was the last straw. They left California, following their family to Siler City.[10]

When immigrants first began to arrive in the mid-1990s, Siler City, like many southern towns, was going through a transition. In the early 1990s, on the eve of the Latino boom, three plants—one furniture and two textile—had closed, in effect laying off five hundred workers. Another five hundred workers lost their jobs in subsequent layoffs, and to make matters worse the state prohibited further construction until the city's sewage treatment plant had been brought up to code. By the fall of 1993 the situation had become sufficiently bad that an article in *Business Week* contrasted Siler City, a backward chicken town, with its high-tech opposite to the north: North Carolina's famous Research Triangle Park. The caricature was a bit unfair. Siler City was still home to upholstery, metal, plastics, and textile plants in addition to furniture and cable makers. New plants soon opened up to fill the gap left by those that had closed.[11]

The *Business Week* caricature did, however, point to a very real problem that small towns throughout the South were facing in the 1980s and 1990s. Without "good" jobs, it was difficult to keep local young people from fleeing to larger cities. There were jobs in Siler City, but few townspeople wanted them. Historic downtowns began to decline. The hardware stores, diners, and small shops that anchored these downtowns pulled out or went bankrupt as box stores, fast food, and the ubiquitous Wal-Mart emerged on the outskirts of the towns. Majestic historic homes still grace downtowns in places like Siler City but now represent a small percentage of the housing stock. Instead, tract homes and trailer parks encircle most southern towns, many of which were losing population prior to the Latino influx. In the case of Siler City, four hundred non-Hispanics left the town during the 1990s.[12]

In this respect, immigrants have saved many southern towns by repopulating them with hardworking families. They provide the backbone of local businesses, churches, and schools. To be sure, this repopulation has had its growing pains. In the case of Siler City, Latinos arrived "overnight." In 1990, Siler

City, with a total population of close to five thousand, was home to only one hundred eighty Hispanics. During the next decade, more than three thousand Hispanics settled in the town, and perhaps as many as ten thousand immigrants arrived in a county that in 1990 had a total population of only forty thousand. Lack of housing, poor medical services, overcrowded schools, and traffic congestion increased tensions. A major drug bust involving Mexican immigrants didn't help matters.[13]

Blanca Hernandez did what many new arrivals do. She got a job working at the Townsend poultry plant cutting and deboning chicken for $250 a week. Blanca's wrists were in pain much of the time, especially when the lines were periodically sped up, but she lasted for about a year. She and Wilfredo had never earned so much money in their lives. She then transferred to a textile mill, where the work was a bit easier, but eventually the repetitive nature of factory labor caught up with her. Blanca was forced onto light duty and had to undergo surgery for carpal tunnel syndrome.[14] Her experience is hardly unique. Pain frequently leads immigrant workers to move from one job or industry to another in the hope of finding work that is less destructive. All too often the search is in vain.

Nevertheless, when the Hernandez family first arrived, Siler City was everything they had hoped for. The work was tough, but it was plentiful, and the cost of living was low.[15] Compared to the rat race of southern California, Siler City was a welcome change. In addition, like townspeople in Mexico and Central America, rural southerners tend to be friendly and polite. Life is tough for early arrivals, but most appreciate "southern hospitality" and have a sense of humor about how little southerners seem to know about Latin-American cultures. As one poultry worker recounts:

When I first came to Arkansas it was tough working in poultry because there were no *Hispanos*. It
was all whites. . . . People in California said, "Oh,
you don't want to go there. They are all racists." . . .
In California, no one was friendly, not the whites,
not the *Hispanos*, and especially not the blacks. Here,
people are friendlier. . . . Still, it was tough. I spoke
little English. And in California they speak different. In the South the accent makes it harder. And
people in California at least know a few words of
Spanish. Here, no. At least not ten years ago. But
slowly I made some friends at work. Really good
people. We never socialized outside of work, though.
Never. Just going out was a struggle. I think people
thought I was retarded. It was like they had never met
someone who didn't speak English. They couldn't
comprehend it. I almost couldn't get my children
registered in school. No one at the school spoke
Spanish! I couldn't believe it. And there was no ESL
[English as a Second Language]. Fortunately, [my
children] spoke a little English, but not much. They
learned quick! What else could they do?

In the beginning, I don't think the whites even
thought about us. There weren't enough *Hispanos*.
If they did they just looked at us like we were from
Mars. I'm serious. Little [white] kids would stare at
us in Wal-Mart. I really felt like a foreigner when I
moved here. I'd been in California for ten years so I
was accustomed to the United States. But here was
like another country.[16]

The reception received by the first immigrants was relatively "warm" in part because many whites either did not no-

tice their new neighbors or saw them as cultural oddities who were just passing through. As one white woman from Arkansas noted:

> I first started to notice them [Latin Americans] in Wal-Mart. You would see a few men, usually in groups of three or four, going around with a shopping cart completely full—like they were shopping for the entire month. It was kind of funny. Then I noticed whole families shopping. The husband, wife, and all the kids shop together. It was really quite adorable. I thought it was the strangest thing at first! Now I don't even give it a second thought. For a long time I never really saw them anywhere except Wal-Mart. Now I see them in more places, but I never really have contact with them.
>
> When I first saw them in Wal-Mart I asked someone and they said they were working in chicken. Those poor people. But it never occurred to me they would stay. I'm not sure if I ever thought about it, really. I just assumed they would work for a bit and then leave. I mean, they're not really from here after all. Right? But then they just kept coming. Before we realized they were here to stay there were so many of them. In some ways it has been good. [*She laughs.*] We got better Mexican food! They are hardworking people. But I know things have been hard in the schools, and with the young men causing trouble. Personally I have never really had much interaction with them.[17]

With few exceptions, whites recall their first interaction with immigrants as having taken place at the local Wal-Mart.

Not church. Not school. Not work. Not the grocery store. Wal-Mart. It is fitting. No two forces have transformed the rural South in the past two decades as much as Wal-Mart and immigration.

The arrival of immigrants has challenged the way southerners understand "their" world. For many, what is often most disturbing is that "by the time we noticed what was happening it was too late. It was like they just appeared. We had no chance to discuss whether it was good or bad, or to prepare for it." For the less politically correct, the feeling can be that "they just came in and took over. They took our jobs, took over our schools, took over our churches. They just took over. It is ruining our culture." As one resident from Siler City put it: "The feeling I got from local officials and government folks at first was that the Latinos were going to come and go. They were seen as migrants. It took a little while for people here to realize they were not going to leave, they were going to stay. Everyone has to adapt to what the city looks like now, and it's a different city from what the older residents grew up in. And these new people are not white Anglo-Saxons, which makes it harder, because after all, this is still the South."[18]

Once residents began to realize that their many new neighbors were here to stay, the reception that immigrants initially received often turned noticeably cooler. Responses varied widely, though in some cases it got ugly. Private security guards at an apartment complex in Nashville, for example, were accused of tormenting the Latino tenants they were in charge of protecting. The guards entered apartments uninvited, handcuffed tenants, held them at gunpoint, and ransacked their belongings. "They kicked residents in the ribs, maced their genitals, and warned, 'I'm going to throw your Spic ass out of the country.'" The owner of the security firm reportedly led the assaults, telling his staff, "I'm bored. Let's go down to taco city and fuck with the Mexicans."[19] Such treatment has been by no means isolated to the southern United States. In Long Island, isolated beatings of migrant laborers have been reported. In Nampo, Idaho, thirty-one high school students were suspended after brawls between whites and Hispanics.[20] Indeed, for all the stereotypes about Mexicans being criminals, the fact is that Latin Americans are more often the victims of crime. Known for avoiding banks, they often carry large sums of money. They tend to walk and are afraid of police, both of which make them even more vulnerable. Wilfredo Hernandez's own father was robbed and assaulted in Siler City.

It is often the case that expressions of anti-immigrant hostility originate from "respectable" citizens. In Burlington, North Carolina, county commissioners unanimously called for a halt to all immigration, legal and illegal. Two-thirds of the residents of ByBee, Tennessee, attempted to block the opening of a Head Start center catering to Latino children. Residents in Lexington, Kentucky, circulated a petition that actually opposed making the city "a safe place for Latinos." According to one poll, almost 80 percent of whites in North Carolina agreed that their neighbors would be opposed to living among Latinos. Similarly, for those without proper papers, it is not uncommon to be arrested for minor crimes such as traffic violations and then deported.[21]

In Cullman, Alabama, Jim Floyd, a sixty-year-old resident, organized an anti-immigration rally that drew over one hundred people. They burned flags of Mexico, the United Nations, and the Communist Party. Floyd wants all Latin-American immigration, legal or otherwise, to stop. "I want my granddaughters to have the same job opportunities as I had, in jobs and culture. I don't want this to be Houston, or Los Angeles. I want it to be Cullman, Alabama. I like my own people more than others, and I am not ready for a world without borders." He has no sympathy for the poultry industry: "If there is a victim here at the moment it's that young Mexican family that all of the sudden is told by Don Tyson: 'Come to America. We'll make you rich.' Those are damn empty promises."[22]

Organized groups such as Project USA and the Federation of American Immigration Reform (FAIR) have fueled these anti-immigration sentiments. In 2000, Project USA erected billboards in eleven states that called immigration a population crisis and cultural problem.[23]

In some places throughout the United States, anti-immigrant sentiment has come from working people who feel their jobs are being threatened by or lost to immigrants. In Mason City, Iowa, 1,400 residents signed a petition to prevent the city from accepting a state grant that would help recruit immigrants in order to alleviate the state's labor shortage. A similar petition emerged in Fort Dodge, about an hour away.[24] The fear of job loss resulting from immigration is not uncommon among working-class whites and blacks, especially in poultry-producing regions (though it varies tremendously depending on location). This is particularly true for working-class African Americans in some regions of the South, many of whom have been unable to make the transition out of poultry and thus are forced to compete with immigrants for low-wage jobs. They complain both about being turned away by plants and about the tendency for line speeds to increase as immigrants make up a larger and larger percentage of plant labor forces. Immigrants, they say, work harder, faster, and without complaining. In addition, some are hired by labor contractors, get paid through a different process, and are thus more difficult to unionize. As a result, racial tensions frequently reach a boiling point.[25]

Nevertheless, many responses to immigration, though sometimes awkward and unsophisticated, are not entirely devoid of good intentions, and may even result from concerns about more overt forms of racism. In Springdale, Arkansas, for example, a group of well-meaning citizens got together in order to educate immigrants about "our culture." As one leader noted:

> Look, can you imagine if we went to Mexico? We wouldn't know what we are doing. Likewise, the Mexicans who come here just need to be educated about our culture. For example, they think it is fine to have five or six adults living in a home. They park their trucks on the front yards and have parties outdoors. You can't do that here. I think if we just tell them, "Hey, this is how we do it here," they will understand. If not, people may get really angry towards Mexicans. We haven't had much of that here, but a little bit. We may also need some new laws about having one nuclear family to a house, or about parking properly. People get worried when it comes to property values. But I think it is really a question of education.[26]

In Siler City, local officials decided to confront the problem directly and formed the "Hispanic Task Force." Unfortunately, the task force did not include a single Hispanic, and its main accomplishment was the publication of a pamphlet warning immigrants that it was illegal to keep chickens and goats within the town limits or to beat up your wife. This is not exactly cutting-edge cultural sensitivity. Unfortunately, the town leaders didn't stop there. The Democratic county commissioner, Rick Givens, became involved in a complaint by police officers about the frequency with which Latin Americans lacked valid driver's licenses or insurance when pulled over in routine traffic stops. Givens cranked off a letter to the INS asking for federal assistance: "More and more of our resources are being siphoned from other pressing needs so that we can provide assistance to immigrants who have little or no possessions. Many of these new needy, we believe, are undocumented or have fraudulent paperwork. We need your help in getting these folks properly documented or routed back to their homes."[27]

Xenophobia aside, the letter points to a real problem. The federal government has done little to help with a variety of complex problems—in schools, on the roads, in hospitals, and so on—that inevitably emerge when large numbers of immigrants arrive in small towns. At the time, Givens saw the issue rather simplistically: "If we have people who are here illegally, why can't we have these people sent home?"[28] That such a policy, if actually implemented, might destroy local factories, small businesses, schools, and churches was never considered.

Neither Givens nor his fellow commissioners could have anticipated the attention the letter would inspire in the national media, or the effect it would have locally. The letter was reprinted in Spanish, causing immigrants to stay in their homes for fear of mass arrests and raids. Even those with legal papers were afraid to go out in public. And although the INS never showed up, the letter seemed to incite individual expressions of anti-immigrant hostility. Immigrants were threatened with deportation by police, public offices became increasingly hostile, and some local businesses even demanded that Latino customers produce identification.[29]

In the fall of 1999, anti-immigrant sentiment in Siler City became more organized when American residents demanded that the board of education address the "problem" of Latinos in the schools. Schools are the lifeblood of small southern communities, and by this date almost 40 percent of students in Siler City were Latinos, with the lower grades especially dominated by Latin-American students. Although Latino students learned English rapidly, English was often their second language. This frustrated the English-speaking students and angered their parents.[30]

In the meantime, however, Givens had visited Mexico on a trip sponsored by the University of North Carolina. By the time he returned, he was a changed man. "I'm going to eat a lot of crow. I still think our government's immigration policy stinks, but I'm not going to make a big deal over legal or illegal. I'm going to help these people acclimate into our community and let the government sort out the work visas."[31] Unfortunately, Givens's change of heart did little to put out the fire he had started in Siler City. A white supremacist group organized an anti-immigration rally featuring David Duke, a leader of the Ku Klux Klan, which in turn drew the national media. Siler City would be the staging ground for future Duke demonstrations across the South and Midwest, none of which were particularly successful. For white supremacists led by Duke, Middle America was the battleground, with anti-immigration the rallying cry.

The night before the white supremacists' rally, St. Julia, a local Catholic church, held a mass in which 250 people— white, black, and Hispanic—prayed that immigrants would find a home in Siler City. The gathering occurred even though the church had been vandalized, tires on a church van had been slashed, and white supremacists had put up signs outside the church saying "White Power," telling immigrants to leave, and proclaiming, "This is our land."[32]

Five hundred people showed up at the Siler City rally led

by Duke, who attacked Givens as a race traitor. Though counter-protesters were encouraged to ignore the rally, a few Latinos showed up to demonstrate. In one meaningful exchange, documented by Barry Yeoman, a Mexican-born Baptist preacher, Israel Tapia, called out "Mr. Duke! Mr. Duke! David!" Duke turned around to see a large Mexican man standing between himself and his admirers. Seizing the moment, Tapia said, "Mr. Duke. Jesus loves you."[33]

The rally was a low point in Siler City race relations. Some Latinos left town, fearful of both white supremacists and the INS. A minority of whites felt empowered to harass their Latin-American neighbors. But most of the residents, white and Latino, went about living their lives. Most whites, even those who were uncomfortable with the presence of Latinos, rejected David Duke and what he represented. And in some instances, the presence of Duke motivated those who had been silent.[34] At St. Julia's, American and Latino parishioners came together in more meaningful ways. As Agnes Bunton, a church member, noted, "After the KKK [came] through town, it just seemed like our parishioners—there were more of them. There was a good reaction in our favor."[35] Blacks, whom supremacists saw as potential anti-immigrant allies, would have nothing to do with Duke and forged church-based alliances with Latinos. For others, such as Givens, the events were something of a learning experience, and, in the end, they tried to do the right thing. A year after the Duke rally, one woman put it this way:

> I never supported Duke, the KKK or any of that stuff. Not even close. Never. But I can't say I really *wanted* Mexicans here either. After the rally I realized they were here to stay whether I liked it or not. I also began to understand how tough it was for

them. They work hard and then have to put up with
that stuff. And they didn't pack up just because this
idiot comes here spouting this crap. I admire them
for that. I think most [Mexicans] understood that
most of us [whites] were better than that [Duke].
After all that, I began to work with some Mexicans
in my church. I felt I had to get to know them. Now
it is different. One friend, Lucila, said she was thinking about leaving Siler City. She wanted to return to
Mexico. I practically begged her not to go. I said
Siler City is your home. I never thought I would say
that to a Mexican. I can't imagine life without Lucila now.[36]

For Latinos, the Duke rally provided a mixed message. The
vast majority remained in Siler City. Companies still needed
their labor, and they still needed jobs. They had too much to
lose to be scared off by the likes of Duke. Many, not surprisingly, chose the most prudent path and did what they had been
doing since arriving in the United States: they kept quiet and
went on with life. For others, the Duke rally was something of
an epiphany.

Duke was scary. California has racism. You see it
every day. But not like that. It was organized and in
your face. It's odd, but after the rally I began to
think of Siler City as my home. I had been here five
years but considered Mexico my home, even though
I knew I would never return. After Duke came I realized this is my home and I better do something
to make it better. I have kids in schools. We go to
church. I work here. It is home. So I started going to

this church that had all kinds of people: Mexicans, Salvadorans, whites, blacks, everyone. Before I went to a Spanish-speaking church. But I said to myself: "We got to stop living in our little worlds." *Hispanos* are just as bad as whites and blacks in this. We come to the U.S. and create our little barrios. And then we wonder why people get pissed. I'm not blaming *Hispanos* for Duke. But we need to do our part.[37]

As a testament to how far immigrants had come since the rally, Latino teens from Siler City launched their own radio program—Grito Juvenil (Teen Shout)—in the summer of 2003. The program, sponsored by a local Latino advocacy group, was put together by some of the town's younger immigrants—the generation that had benefited from their parents' hard work in local poultry plants and now had the confidence and English to participate in community life. One of the program's hosts, Nayeli, had been born in Mexico, arrived with her parents in the 1990s, was now on the high school soccer team and honor roll, and was the only Hispanic member of an otherwise African-American gospel choir.[38]

Siler City did not become a multicultural paradise. But the Duke rally did jolt residents and continues to serve as a warning to those of us who feel uncomfortable with the xenophobia represented by the likes of David Duke, Patrick Buchanan, Rush Limbaugh, or Bill O'Reilly. What Siler City tells us is that the actions of local political, community, and church leaders are particularly meaningful. They can, in quite real ways, not only shape the reception that immigrants receive, but determine to a great degree the subsequent nature of race relations. By writing his letter to the INS, Givens chose a path that got ugly and out of control in a remarkably short time.

Fortunately, more hopeful examples abound. While it is often the worst cases that make the headlines, most towns are finding new ways of getting along. In Perry, Iowa, for example, home to a large IBP beef processing plant, town leaders intervened in order to stop racial tensions from growing worse. They quickly brought in Spanish-speaking teachers and a bilingual police officer while forming a diversity committee. In Dalton, Georgia, community leaders recruited teachers from Mexico to teach bilingual classes.[39] The results, though modest, contrast sharply with the situation in Siler City.[40] And one of the most concerted and well-studied efforts in dealing with newcomers has been in Garden City, Kansas, a small town in America's heartland with a strong meatpacking industry that has attracted large numbers of immigrants. The town's relative success demonstrates that through hard work and forethought towns can adjust to and benefit from the changes brought by immigration. Donald Stull and Michael Broadway, researchers who both studied and played a positive role in the town's transition, put it this way:

> Clergy, educators, social service providers, law enforcement officials, and local journalists have struggled to provide a positive context of reception for Garden City's newcomers, whatever their backgrounds. It has not been easy. Two decades of sustained growth have combined with high population mobility and dramatically increasing ethnic and linguistic diversity to present what are remarkable challenges: housing shortages, soaring school enrollments, rising rates of crime and social problems, insufficient medical services, and an overburdened road system. But Garden City has met these chal-

lenges head on and emerged as an exemplar for cities and towns throughout North America that are facing rural industrialization, rapid growth, and increasing ethnic and linguistic diversity.[41]

More hopeful examples notwithstanding, Siler City highlights the difficult position in which small-town America finds itself. Enlightened leaders can ease the transition, but with immigration comes considerable growing pains, especially in the beginning and especially in small towns with few resources. Complaints from small-town leaders, even when they are expressed in rather impolite ways, need to be addressed rather than dismissed or ridiculed.

Unfortunately, a decade into the regionwide influx of Latinos, only one southern state, North Carolina, has an advisory board on Hispanic affairs.[42] Moreover, the absence of federal assistance to help small towns and rural counties with the adjustments brought by immigration is nothing short of disgraceful. Providing services to immigrants has not been a particularly popular way of spending money in the U.S. Congress. The very same politicians who love to bash immigrants for political purposes quietly facilitate their entry on behalf of the corporations who depend on their labor. They then do little to help the communities receiving vast numbers of immigrants. Businesses, perhaps predictably, are not any better. They profit off the labor of immigrants but do little to help them or their U.S. communities deal with the transition. As a larger society, we ignore these issues at our peril. By leaving communities like Siler City to fend for themselves, we risk inflaming racial prejudices and hampering the development of diverse, healthy communities.

VIII
Toward a
Friendlier Chicken

These days, read any description of how chickens go from
downy hatchlings to lunch salads and roasted dinner entrees
and you'd swear that someone had slipped you the script for
an episode of X Files or the latest Stephen King thriller,
Poultrygeist. All the ingredients for a devilish tale are there:
epidemics of Salmonella stalking unsuspecting consumers;
slaughterhouse workers toiling in ghoulish conditions;
stomach-wrenching mountains of manure and chicken
carcasses; and brutally overcrowded factory farms.
Trouble is, none of this is fictional.

hicken's postwar success was due in large part to its
affordability and healthfulness. Recently, however,
the safety and health of American chicken has
come under increased scrutiny. This attention has
put industry leaders on the defensive, but it has also created

opportunities. Tyson, for its part, has recently jumped on the organic bandwagon with its Nature's Farm Organic Chicken. With an "if you can't beat 'em, join 'em" attitude, the company proclaims on its Web site: "Mother Nature has been growing chickens a lot longer than we have. So we decided to follow her example and go back to the simpler, more natural way of doing things. The result is healthy, naturally-good chicken that's so delicious you'll be amazed." These chickens are raised on an "all-natural ... vegetable diet of organically grown grains (that means no pesticides!), plus a blend of the natural vitamins and minerals they need (just like you) to grow up strong and healthy." Unlike the typical industrial bird, Tyson's organic chickens have "room to spread their wings" before they are slaughtered and placed on an "environmentally friendly recyclable" tray.[1]

That the leading producers of industrial chicken can no longer afford to dismiss "organic" and "free-range" as crazy ideas is instructive. It reflects the growing concern about the healthfulness of chicken and other foods we consume. But our definition of "health" needs to continue to expand. The health of chicken is not determined solely by what the bird eats or whether it gets to stretch its legs before being slaughtered. Nor can health be reduced to the fat, cholesterol, and protein content of chicken. The medicines and chemicals that the food industry puts into our food are an important part of its healthfulness, but our concern should not end there. The health of food workers and farmers must also be considered as well— not simply because they are getting an unfair deal, but also because they are, I believe, the people who are most likely to watch out for consumer interests. If we do not want to return to the farm and grow our own food, we need to empower workers and farmers. Only by improving their lives and giving them more control over the foods they produce can we improve the lives of everyone along the chain of food production and consumption.

The good news is that more and more groups, from environmental organizations and churches to consumer advocates and labor unions, are taking a hard look at industrial agriculture by exploring the connections among worker safety, farmer well-being, consumer health, and environmental concerns. This critical examination of agribusiness has, in turn, led groups of farmers, workers, consumers, and environmentalists to explore alternative ways of producing food.

The Sierra Club Versus Industrial Chicken

Perhaps no group has spent more time and effort taking on big chicken (and "factory farming" in general) than the Sierra Club. At first glance, this matchup may seem a bit odd. With some seven hundred thousand members, this "club" is a mainstream environmental organization that has traditionally worked to save forests, wetlands, jungles, and wildlands while stopping commercial logging, rainforest destruction, and suburban sprawl. Its methods are varied, ranging from education and lobbying to political protest and legal action.

Yet the Sierra Club's interest in industrial agriculture should not be entirely surprising. Farms—particularly concentrated animal feeding operations (CAFOs) that raise cattle, hogs, and poultry in confined places—have replaced factories as the leading polluter of America's waterways.[2] Moreover, the Sierra Club's environmental focus has led it to a broader set of related concerns regarding workers, farmers, and consumers. The interconnectedness of work, consumption, and the environment, so central to agriculture today, is highlighted in the

Sierra Club's *Rap Sheet on Animal Factories*. The *Rap Sheet* is an accounting of legal cases against agribusiness's premier corporations. It is not a complete list of corporate legal troubles; it excludes, among others, those violations that failed to make it into the legal record. Nevertheless, the *Rap Sheet* suggests that the industry leaders—Tyson, ConAgra, Gold Kist, Cargill, Cagle's, Foster—spend a lot of time in court. Indeed, the quantity and tenor of the cases suggest that breaking the law is routine.[3] Take ConAgra, Tyson's rival in the bid to acquire Holly Farms. With operations in twenty-five countries, ConAgra is the second largest food company in the United States. Some of its meat brands include Armour, Butterball, Country Pride, Eckrich, and Hebrew National, but its empire also includes everyday food products such as Banquet, Bumble Bee, Chef Boyardee, Healthy Choice, Knott's Berry Farm, and Orville Redenbacher. In Carthage, Missouri, ConAgra owns a turkey plant that slaughters approximately thirty thousand birds a day and discharges about 1.3 million tons of treated slaughterhouse waste per day. Since 1990, regulators have recorded dozens of violations of pollution limits. In one case, one of the processing plant's lagoons was found to be leaking nearly a million gallons of waste per month. Once caught, the company continued to use the lagoon for four more years, during which time it received six written requests from Missouri state officials asking for a plan to close the lagoon. Why close it? The company had been fined only $42,000, a drop in the bucket compared to the $2.8 million worth of products that the plant sold during that period to the federal food assistance programs, including the school lunch program.

As the *Rap Sheet* points out, ConAgra is not alone in breaking the law. According to the Environmental Protection Agency (EPA), Foster Farms, the largest poultry company in the West and a top-ten producer nationally, pled guilty in 1998 to "negligently discharging approximately 11 million gallons of storm water polluted with decomposed chicken manure into the San Luis National Wildlife Refuge in violation of the Clean Water Act." The company agreed to pay a criminal fine of $500,000.

In January 2000, Central Industries, along with its five subsidiaries (including BCR Foods, a top-twenty poultry producer), were indicted in Jackson, Mississippi, for violating the Clean Water Act. The company processes thousands of tons of poultry by-products, some of which, including untreated blood, bypassed the rendering plant and went directly into a wastewater lagoon, eventually seeping into local rivers. The company violated its discharge permit more than 1,100 times by depositing unacceptable levels of pollutants into the Shockaloo Creek, a tributary of the Pearl River, which supplies drinking water for the city of Jackson.

Simmons Food, a top-twenty producer, has attracted the attention of two states. Its plant in Southwest City, Missouri, has had "chronic environmental problems for years" and is currently operating under consent decrees in both Missouri and Oklahoma. It routinely violated an earlier consent decree and racked up fines in both states totaling close to $1 million. In fact, since 1987 the Missouri Department of Natural Resources has issued the plant twenty citations for "chronic noncompliance" involving environmental violations.

What is notable about these violations is their chronic and willful nature. The companies know they are violating the law and continue to do so even after they are caught red-handed. Unfortunately, as a 2005 Human Rights Watch report makes clear, the industry's lack of concern for the environment is matched by its poor treatment of workers. The fact that Human Rights Watch, an organization that traditionally investigates "other" places in the world, thought it necessary to examine the U.S. meat and poultry industry for human rights violations is telling. It summarized its findings in blunt terms: "Constant fear and risk is . . . a feature of meat and poultry labor. Workers risk losing their jobs when they exercise their rights to organize and bargain collectively in an attempt to improve working conditions. Employers put workers at predictable risk of serious physical injury even though the means to avoid such injury are known and feasible. They frustrate workers' efforts to obtain compensation for workplace injuries when they occur. They crush workers' self-organizing efforts and rights of association. They exploit the perceived vulnerability of a predominantly immigrant labor force in many of their work sites. These are not occasional lapses by employers paying insufficient attention to modern human resources management policies. These are systematic human rights violations embedded in meat and poultry industry employment."[4]

The finding of Human Rights Watch are echoed in the Sierra Club's *Rap Sheet*. In 1999, for example, the Occupational Safety and Health Administration (OSHA) cited two poultry-processing plants owned by BCR Foods for workplace safety violations. One offense resulted in a worker being electrocuted. In another case, launched in 1997, OSHA cited Cagle's Inc., a top-ten poultry producer out of Atlanta, for numerous "willful and serious safety violations" at its Macon, Georgia, facility. According to OSHA, its "action comes after an employee lost part of a finger while cleaning moving equipment on September 10, 1996, and two other workers suffered amputations of a finger and a foot the preceding year." OSHA's investigation demonstrated that even after two workers lost their limbs, the company had failed to take the necessary precautions. As a result, another worker was maimed.

Hudson Foods, which was purchased by Tyson in 1998, was cited a year before the takeover for worker safety violations. OSHA noted that since 1973 its investigators had visited the plant twenty-three times and issued citations on sixteen of those visits. In the 1997 visits, OSHA reported that "hazards found . . . resulted in more than three hundred cases of cumulative trauma disorders . . . among workers." Several years later, with Tyson in control, the company was forced to pay a $20,000 fine for failing to immediately report four "caustic releases of airborne ammonia . . . [that] cause[d] employee injuries and evacuation of the general public."

In a particularly egregious case, IBP, now owned by Tyson Foods, was ordered to pay $10.8 million to an employee who was wrongfully discharged. The employee was fired when he refused to illegally deceive his stepson into signing an injury settlement and waiver after the son had lost all of the fingers in his right hand in a workplace accident. IBP, apparently, could not imagine that one of its employees would not turn on his own son.

One of the most outrageous abuses belongs to none other than Tyson Foods. In October 1999, the U.S. Labor Department levied the maximum possible penalty against Tyson, a mere $59,274, for two violations of the child labor provisions of the Fair Labor Standards Act. The sad irony is that the only reason the company got caught for hiring children was that the underage employees were involved in accidents. The first, a fifteen-year-old, was electrocuted when he bumped into a fan while working as a chicken catcher. The second youth was maimed while working illegally at a plant in Missouri.

Considering that plant work is dangerous and low-paying, one might expect companies to pay their employees according to the law. Unfortunately, this is not the case. A 1997 Labor Department study found that fewer than 40 percent of all poultry-

processing plant operations were in compliance with the federal Fair Labor Standards Act. In other words, most plants were
routinely violating workers' rights. Perhaps not surprisingly, a
class-action suit has been ongoing for years on behalf of poultry workers throughout the country who have been routinely
denied pay for the time they spend putting on, sanitizing, and
taking off protective clothing.

Another common violation is the failure of poultry companies to pay overtime to chicken-catching crews, a sad irony
given that these are among the worst jobs in the industry.
Catchers go from farm to farm rounding up fully grown chickens in the middle of the night, stuffing them into cages, and
loading them onto trucks for delivery to processing plants. The
work is brutal. Workers wear little protective clothing as they
stoop down to pick up six or seven panicked birds by their
powerful claws. The work often cannot be completed in an
eight-hour shift, but companies nonetheless fail to pay overtime. Perdue, for example, was forced to compensate 177 such
workers a total of $300,000 in unpaid overtime. Wayne Farms,
the sixth-largest poultry producer in the country, paid almost
$150,000 to thirty-seven workers for the same violation. As a
result of these and other problems, integrators are now hiring
independent contractors to do the work, effectively relieving
the industry of any responsibility for pay, working conditions,
benefits, vacation time, and the like.

Unfortunately, the industry record is equally troubling
when it comes to consumer safety. Contaminated chicken kills
at least one thousand people and sickens millions of others
every year in the United States. *Time* magazine calls chicken
"one of the most dangerous items in the American home," and
government officials suggest we simply assume poultry contains lethal microbes. A report by the Center for Science in the

Public Interest found that approximately 30 percent of chicken is contaminated with *Salmonella,* and 62 percent with *Campylobacter.* According to the USDA, these two pathogens are responsible for 80 percent of the illnesses and three-quarters of the deaths connected to meat consumption.[5]

Modern factory farms are efficient in many ways, including their capacity to cultivate and spread diseases. Likewise, processing plants do an excellent job of moving pathogens from one bird to the next. Dead chickens, for example, are given communal baths in a toxic brew popularly known as "fecal soup." The bird that emerges from these soup tanks is, according to a former USDA microbiologist, "no different than if you stick it in the toilet and ate it."[6]

Worse yet, government oversight is inadequate. The USDA has a long history of siding with the poultry industry against consumer safety. While European governmental organizations have slowed line speeds and improved regulatory oversight, the USDA has maintained an archaic system of inspection that has allowed the industry to steadily increase line speeds and reduce health standards. According to the General Accounting Office, an investigative wing of Congress, the current inspection system is "only marginally better than it was 87 years ago when it was put in place." In fact, during the Reagan era of deregulation, the USDA drastically cut its already measly inspection staff, which caused over one thousand inspector positions to be vacant by the late 1990s.[7] As a result, according to a government report:

> Up to one-quarter of slaughtered chickens on the inspection line are covered with feces, bile, and feed. Dead and diseased animals are processed and end up in the supermarket.

Chickens are soaked in baths of chlorine to remove
slime and odor.

Mixtures of excrement, blood, oil, grease, rust, paint,
insecticides, and rodent droppings accumulate
in processing plants.

Maggots and other larvae breed in storage and transportation containers, on the floor, and in processing equipment and packaging, and they drop
onto the conveyor belt from infested meat splat-
tered on the ceiling.

Slaughterhouses—which by law must be inspected
once every shift—go as long as two weeks without inspection.[8]

As the *Rap Sheet* suggests, conditions in individual plants
reflect this general threat to consumer health. In 2000, Cargill,
the nation's largest privately owned corporation, distributed
bacteria-contaminated turkey products from its Waco, Texas,
plant, causing the deaths of four people, twenty-eight cases of
listeriosis, and three miscarriages or stillbirths.

Cargill is not alone. In October 2001, Pilgrim's Pride,
a top-four poultry producer, recalled almost thirty million
pounds of frozen ready-to-eat poultry products and closed its
plant in Franconia, Pennsylvania, when it learned that it was
the likely source for the deadly *Listeria monocytogenes.* By the
time the source of the poultry was definitively determined,
fifty people had become ill. Most were hospitalized, seven died,
and three had miscarriages or stillbirths. The source of the
problem was most likely precooked turkey deli meats, and the
Centers for Disease Control and Prevention found the *Listeria*
bacteria in two plants. That Pilgrim's Franconia plant was one
of them should have come as no surprise to investigators. Prior

to the outbreak, the plant had received forty citations for sanitation violations in less than a year, ten of which had been issued in the month prior to the incident.

Terms like "chronic," "willful," and "repeat violations" used in the citations suggest that giant food companies know what they are doing and frequently continue to engage in criminal behavior even after being caught—behavior that threatens or destroys workers, farmers, the environment, and consumers.

What, then, can be done to reform big chicken? Mounting legal cases against violators—as the Sierra Club is doing with great success—is one way that corporations can be held accountable. To be sure, the obstacles are considerable. The poultry industry includes some of the largest companies in the world. With their immense resources, they are more than willing to engage in costly legal battles—and pass along the costs of any legal sanctions to the consumer.

As a result, more and more groups, from environmentalists and labor unions to animal rights activists and farming advocates, are pressuring the government to enforce and strengthen existing regulations (and stiffen fines). They, too, face considerable obstacles. Agribusiness remains a powerful lobby, especially on the state level, and in fact the government has been part of the problem for the past century. Government regulation and inspection may limit the industry's worst transgressions, but by and large the state has looked the other way and even helped institutionalize the current way we process meat and poultry.

Despite these obstacles, however, the government should be pressured to:

> Fully implement and enhance the existing USDA
> system of plant inspections to comprehensively

monitor sanitation and guarantee the safety of our food. In short, put some teeth back into the current USDA inspection system.

Empower a state agency to monitor working conditions in meat-processing plants. USDA inspectors currently roam plants to monitor food safety; worker safety needs at least as much vigilance.

Reduce line speeds to a rate that is manageable for workers and allows them to play a serious role in monitoring the safety of the foods they are processing.

Enforce and strengthen pollution regulations regarding waste disposal and water treatment at plants, and make integrators responsible for the disposal of waste on farms.

Impose serious penalties on repeat offenders. Close plants that routinely violate the law.

Provide financial and other incentives to encourage companies to produce a healthier bird, improve working conditions, use fewer resources, treat farmers fairly, and respect the environment.

Establish a system of product labeling that tells consumers exactly what they are getting when they purchase chicken, beef, and pork in stores and restaurants. This should include information about nutrition, safety and sanitation standards, working conditions, and even the environment.

Above all, the government must level the playing field for farmers and workers. Farmers need to be able to negotiate their contracts in good faith. They need to get paid in a way that is not only transparent, but also reflects the massive investments they make in the industry. They also need to operate without fear of retaliation from integrators. Workers, regardless of where they are born, must labor under safe conditions, get paid a living wage, and not fear their employers. This will require serious immigration reform. The government—pushed by consumers, workers, farmers, environmentalists, and others—has to help create and support a legal and political climate in which workers and farmers can organize.

Friendly Chicken

It is difficult to be optimistic about an industry in which power is so concentrated, government involvement so one-sided, and abuses so routine and outrageous. There is something fundamentally wrong with an industry in which the very people who produce chicken are punished when they identify problems with food safety, working conditions, or farming practices. The existing system needs to be reformed, but we also need to look for alternatives (which, through competition, can push agribusiness to initiate reforms).

There are indeed reasons to be hopeful. As we have seen, during the 1980s and 1990s the poultry market became highly differentiated as thousands of different chicken products were introduced. The price of chicken was so low that consumers were willing to pay more for better, or at least different, products (as they have been for coffee, beer, water, and so on). Many of these poultry products are, unfortunately, unhealthy. But it doesn't have to be that way. The continual creation of new poultry products has raised the possibility of doing chicken in entirely new (and profitable) ways. Millions of Americans are already paying 15 to 30 percent more for chicken that is "air-chilled," 50 to 75 percent more for meat that is free of antibiotics and hormones, and more than double for kosher, free-range, and organic chicken. Demand for chicken continues to climb, particularly for niche products. People are willing to spend a little more for chicken treated with higher standards.[9]

In general, the premium or niche chicken that is on the market today is limited in two basic respects. First, the high cost, particularly of organic and free-range chicken, largely restricts these products to natural foods stores and their middle- and upper-class customers. Yet the phenomenal growth of these stores, such as Whole Foods and Wild Oats, suggests that the market for these foods will continue to expand and further penetrate mainstream supermarkets. There is also good reason to think that the price of organic chicken will continue to decline as alternative feeds become more readily available. At least in the short to medium term, however, these products will occupy a small niche in a market otherwise dominated by the industry leaders. We need a bird that is both better for us and more affordable.

Second, and more importantly, despite thousands of different poultry products, none of the chicken on the American market today addresses the broader concerns of worker and farmer welfare and empowerment. Indeed, it is all too easy to produce "healthy" organic and free-range chicken in a way that differs very little from industrial chicken in terms of the highly exploitative and inequitable working conditions under which it is produced. This is precisely why companies such as Tyson have moved so quickly into organic chicken. It requires very little change in the way that industry leaders do business.

In this respect, the most promising alternative product is "Friendly Chicken." The basic idea behind Friendly Chicken is to deliver an economical product line that will appeal to the widest array of consumers in a way that is consistent with principles of equity, social justice, and environmental sustainability. This is a natural chicken free of hormones, antibiotics, and other additives. A Friendly Chicken tastes better, is healthier, and is grown and processed in a manner that actively maintains high labor and environmental standards. It is also marketed and distributed in a fundamentally different way from industrial chicken.

The company that is creating this product, Bay Friendly Chicken, was incorporated only in April 2004, but it has been developing for close to a decade. It emerged with the active support of the Chesapeake Bay Foundation and the Delmarva Poultry Justice Alliance in conjunction with growers, catchers, plant workers, business analysts, and others.[10] More recently, it has benefited from a grant from the USDA. That Bay Friendly Chicken has emerged in the industry's birthplace—Delmarva— is perhaps only fitting, although the basic tenets of Friendly Chicken depart fundamentally from existing models of poultry production:

> Local control. Vested owners of the company must live in the area where the company operates, and no single owner may hold more than 5 percent of the company's stock value.
>
> Shared decision-making. In addition to shareholders, the company's board of directors must include environmentalists, growers, workers, and others from the industry and local community. The board must have not only superb business skills, but also integrity, diversity, and a responsibility to the local community, industry workers, and consumers.
>
> A better bird. Chickens are given more space, more ventilation, more natural lighting, more frequent litter clean-out, more growing time, and more

humane treatment. The company also pledges to avoid the unwarranted and indiscriminate use of hormones and antibiotics and to keep their products free from preservatives.

Fair treatment of growers. Growers should be treated fairly, be part-owners in the company itself, and receive a higher percentage of the market price. They should also retain control over their farms as long as their standards are consistent with the delivery of high-quality chickens.

High environmental standards. The company will help growers adopt more environmentally sound practices and employ technologies that reduce the environmental impact of processing plants.

High labor standards. All workers associated with the production and processing of chickens must be paid a living wage, receive adequate health insurance and retirement benefits, and maintain the right to collective bargaining. Workers should also be given an ownership stake in the company.

Food safety. Slower line speeds and happier workers are needed to ensure quality control, and new technologies and more exacting inspections are required to ensure food safety. In particular, "air-chilling" should replace the current industry practice of water chilling (whereby chickens are cooled in communal baths in which microbes spread from one bird to the next). This change would also improve the taste of chicken because water-chilling the bird leaches much of its flavor.

Sound fantastic? Perhaps. These commitments and values do increase the cost of chicken, but a healthier, better-tasting chicken with real value added would also inspire consumers to pay a higher price, as they have in the past for other improvements in quality. And the additional cost is not as much as one might think. According to Bay Friendly Chicken, paying workers and catchers a living wage, for example, will cost the consumer about four cents a pound and result in a labor force that performs better under less supervision and with less training, is absent from work less often, has higher morale, and is healthier. Most importantly, this labor force, which would hold partial ownership in the company, will be encouraged to identify where the chicken is not meeting food safety and environmental standards. All in all, the Friendly Chicken would cost consumers about 50 percent more, raising the cost of chicken from roughly $1 a pound to about $1.50 a pound.

In the long run, as the market develops, it is quite possible that the price of Friendly Chicken could decrease. More genetically diverse breeds of chicken, with stronger immune systems, would require fewer antibiotics and medications to survive the grow-out period. Workers would be more productive. Farmers would be better rewarded. And consumers would get a bird that is more than a marketing ploy; it would represent a fundamentally different way of producing food. The obstacles to supporting this type of company in an industry dominated by large agribusiness may be significant, but they are worth overcoming. It is time to reclaim poultry's potential, hold Herbert Hoover to his promise, and put a Friendly Chicken in every pot.

Notes

Introduction

1. "Who's to Blame? Obesity in America: How to Get Fat Without Really Trying," *ABC World News Tonight with Peter Jennings* (Dec. 8, 2003); see http://abcnews.go.com/sections/WNT/ Living/obesity_031208–1.html. I chose ABC News, but could have turned to any of the major media outlets to make the broader point about our current preoccupation with food and fat.

2. Ibid.

3. "Obesity Gains on Smoking as Top Cause of U.S. Death," abcnews .com (Mar. 9, 2004).

4. Ibid.

5. Bruce Horowitz, "Eating Choices: Junk Food or Healthy Food," *USA Today*, July 9, 2003; see www.baxterbulletin.com/news/stories/20030709/local news/614699.html.

6. Anne Pleshette Murphy, "Bombarded by Food Ads," abcnews.com (Dec. 2, 2003).

7. Geraldine Sealey, "Beyond Baby Fat," abcnews.com (Sept. 30, 2003).

8. There are some excellent books on the role of industry and other factors in influencing the foods we eat. See, for example, Marion Nestle, *Food Politics: How the Food Industry Influences Nutrition and Health* (University of California Press, 2002); Kelly D. Brownell and Katherine Battle Horgen, *Food Fight: The Inside Story of the Food Industry, America's Obesity Crisis, and What We Can Do About It* (McGraw-Hill, 2003); and Greg Critser, *Fat Land: How Americans Became the Fattest People in the World* (Mariner Books, 2004).

9. Nestle, *Food Politics*, p. 22.

10. "Who's to Blame?"

11. Ibid.

12. Ibid.

13. Ibid. See Nestle's *Food Politics* for a much more thorough discussion of the relationships among the federal government, the food industry, and our health.

14. Getting this type of information is possible in our information age, even if it requires a little digging. For example, there is a wonderful study that tells you everything you wanted to know about tomatoes (and much more): Deborah Barndt, *Tangled Routes: Women, Work, and Globalization on the Tomato Trail* (Garamond Press, 2002). Jane Dixon's look at chicken in the Australian context is also in this vein: see *The Changing Chicken: Chooks, Cooks, and Culinary Culture* (University of New South Wales Press, 2002).

15. Phil Lempert, "The Scoop on Cereals," *Supermarket Guru* (Mar. 14, 2001), see www.supermarketguru.com/page.cfm/286; www.nsda.org.

16. Kari Lyderson, "Fowl Behavior," *In These Times*, Mar. 19, 2001, p. 8; Russell Cobb, "The Chicken Hangers," *Identify*, Feb. 2, 2004.

17. Lyderson, "Fowl Behavior," p. 8.

18. It's not hard to imagine book-length studies on any number of chicken-related topics, from consumer preferences, scientific advancements, and worker recruitment to industry consolidation, product development, and the raising of chickens. Such studies have their virtues, and where available I have drawn from them. Ultimately, however, I decided to paint this poultry portrait in fairly broad strokes, and to focus on connections.

19. The classic ethnographic study of factory labor is Michael Buroway, *Manufacturing Consent: Changes in the Labor Process Under Monopoly Capitalism* (University of Chicago Press, 1979). Other important "early" works include Maria Patricia Fernandez-Kelly, *For We Are Sold, I and My People: Women and Industry in Mexico's Frontier* (State University of New York Press, 1983); and William E. Thompson, "Hanging Tongues: A Sociological Encounter with the Assembly Line," *Qualitative Sociology* 6, no. 3 (1983): 215–237. For an excellent, more recent look at factory work, see Kevin A. Yelvington, *Producing Power: Ethnicity, Gender, and Class in a Caribbean Workplace* (Temple University Press, 1995). The obvious starting place for accounts of the meat industry is Upton Sinclair, *The Jungle* (1906; Barnes and Noble Books, 1995). For a thorough and recent firsthand account inside a meatpacking plant, see Deborah Fink, *Cutting into the Meatpacking Line: Workers and Change in the Rural Midwest* (University of North Carolina Press, 1998). Some excellent accounts from journalists include Tony Horowitz, "9 to Nowhere," *Wall Street Journal*, Dec. 1, 1991, p. A1; and Charlie LeDuff, "At a Slaughterhouse, Some Things Never Die," *New York Times*, June 16, 2000.

Chapter 1
Love That Chicken!

1. We will return to the science of chickens in Chapter 2. For more detailed treatments see William Boyd, "Making Meat: Science, Technology, and American Poultry Production," *Technology and Culture* 42 (Oct. 2001): 631–664; William Boyd and Michael Watts, "Agro-Industrial Just-in-Time: The Chicken Industry and Postwar American Capitalism," in David Goodman and Michael Watts, eds., *Globalising Food: Agrarian Questions and Global Restructuring* (Routledge, 1997).

2. This history has been traced by a number of authors. For a wonderfully readable and condensed version, see John Steele Gordon, "The Chicken Story," *American Heritage* (Sept. 1996): 52–67.

3. Chul-Kyoo Kim and James Curry, "Fordism, Flexible Specialization and Agri-Industrial Restructuring," *Sociologia Ruralis* 33, no. 1 (1993): 67.

4. *PR Newswire,* May 11, 1984.

5. "McDonald's Hopes to Rule the Roost with New Chicken Promotion," *Crain's Chicago Business,* Oct. 2, 1995, p. 35.

6. N. R. Kleinfield, "America Goes Chicken Crazy," *New York Times,* Dec. 9, 1984, sec. 3, p. 1.

7. "Fresh Chicken," *Consumer Reports* 54, no.2 (Feb. 1989): 75.

8. Ibid.

9. "Dueling for the Center of the Plate," *National Provisioner* 210, no. 3 (Mar. 1996): 54; Gal Group, Inc., "Boom in Chicken Consumption Has Room to Grow," *Refrigerated and Frozen Foods* 11, no. 6 (June 2000): SOI–53.

10. "September Is National Chicken Month," *PR Newswire,* Sept. 10, 2001. Precise statistics can be tricky; they depend on who is doing the measuring, whether one is interested in production or consumption, and whether the numbers are compared in terms of value, weight, or some other criterion. According to industry data, per capita chicken consumption was just under eighty pounds, with beef just under seventy, and pork a little over fifty. See Donald D. Stull and Michael J. Broadway, *Slaughterhouse Blues: The Meat and Poultry Industry in North America* (Wadsworth, 2004). Regardless of numbers, however, it is clear that chicken consumption soared during the postwar period and around 1990 overtook beef, which started declining in the mid-1970s as America's favorite meat. The "Poultry Yearbook" put out by the Economic Research Service of the USDA provides a sense of how many different ways statistics on poultry can be calculated. For a very readable, if slightly dated, government publication that charts the rise of poultry in relation to other meats, see Floyd A. Lasley et al., *The U.S. Broiler Industry* (USDA-ERS, 1988).

11. Kleinfield, "America Goes Chicken Crazy," p. 1.

12. Ibid.; Kim and Curry, "Fordism," pp. 65–70; Gordon Sawyer, *The Agribusiness Poultry Industry: A History of Its Development* (Exposition Press, 1971); Boyd and Watts, "Agro-Industrial Just-in-Time," pp. 192–225.

13. Kleinfield, "America Goes Chicken Crazy," p. 1. To understand the rise of processed foods as a business strategy, see Marvin Schwartz, *Tyson: From Farm to Market* (University of Arkansas, 1991).

14. Kleinfield, "America Goes Chicken Crazy," p. 1.

15. Schwartz, *Tyson,* chap. 1.

16. "Putting Its Brand on High-Margin Poultry Products," *Business Week,* Aug. 20, 1979, p. 48.

17. Laura Konrad Jereski, "The Wishbone Offense: Branding a Commodity," *Marketing and Media Decisions* (May 1985): 80–84.

18. Dick Anderson, "Don Tyson Rules the Roost," *Southpoint* (Nov. 1989): 16–20.

19. *1987 Annual Report,* Tyson Foods, Inc.

20. Kleinfield, "America Goes Chicken Crazy," p. 1.

21. Ibid.

22. Anderson, "Don Tyson Rules the Roost," p. 19.

23. "Perdue: He Was the First to Brand the Chicken Successfully," *Media Decisions* (Apr. 1974): 62, 112–116.

24. Ibid., pp. 112–116.

25. Ibid.

26. Ibid.

27. Nancy Giges, "Ad Outlays Become Important as Brand Names Gain in Chicken Market," *Advertising Age* (Sept. 13, 1976): 50–54.

28. Joe G. Thomas and J. M. Koonce, "Differentiating a Commodity: Lessons from Tyson Foods," *Planning Review* (Sept.–Oct. 1989): 24–29.

29. "Fresh Chicken," p. 75; John Robbins, *Diet for a New America* (Dimensions, 1987).

30. "Fresh Chicken," p. 75.

31. Giges, "Ad Outlays," pp. 50–54.

32. Jereski, "Wishbone Offense," p. 84; *1994 Annual Report,* Tyson Foods, Inc.

33. Julie Liesse-Erickson, "Meat Marketers Ply Branding Iron," *Advertising Age* (May 29, 1989): 28.

34. Jereski, "Wishbone Offense," p. 84; "Fresh Chicken," p. 75.

35. Kleinfield, "America Goes Chicken Crazy," p. 1.

36. Laura Konrad Jereski, "Fast Palate-Pleasers Tempt Baby Boomers," *Marketing and Media Decisions* (Jan. 1985): 83–90.

37. Eric Schlosser, *Fast Food Nation: The Dark Side of the All-American Meal* (Houghton Mifflin, 2001), p. 139.

38. Christine Dugas and Paula Dwyer, "Deceptive Ads: The FTC's Laissez-Faire Approach Is Backfiring," *Business Week,* Dec. 2, 1985, p. 136.

39. "September Is National Chicken Month."

40. Jereski, "Fast Palate-Pleasers," p. 88.

41. In 1985, for example, Holly Farms had Dinah Shore promote its nuggets in a $6-million advertising campaign. See Jereski, "Wishbone Offense," p. 84.

42. Bradley H. Gendell, "Cockfight," *Financial World,* July 11, 1989, pp. 26–27.

43. Kleinfield, "America Goes Chicken Crazy," p. 1.

44. Gale Group, Inc., "2000 State of the Industry Report: Consumption Measurement," *National Provisioner* 214, no. 8 (Sept. 2000): 55.

45. "Chicken Is La King," *Foodservice Director* 13, no. 8 (Aug. 15, 2000): 91.

46. Laura A. Brandt, "Unleashing Poultry's Potential," *Prepared Foods* 169, no. 1 (Jan. 2000): 36.

47. James S. Eales and Laurian J. Unnevehr, "Demand for Beef and Chicken Products: Separability and Structural Change," *American Journal of Agricultural Economics* (Aug. 1988): 521–532.

48. "Fresh Chicken," p. 75.

49. Susan E. Gebhardt and Robin G. Thomas, *Nutritive Value of Foods* (U.S. Department of Agriculture, 2000).

50. Jayne Hurley and Bonnie Liebman, "Fast Food Follow Up: What's Left to Eat?" *Nutrition Action Healthletter* (Nov. 1997).

51. Bonnie Liebman and Jayne Hurley, "Fast-Food 2002: The Best and Worst," *Nutrition Action Healthletter* (Sept. 2002).

52. Ibid.

53. "Heavy Kids' Menus," abcnews.com (Feb. 24, 2004).

Chapter 2
An American Industry

1. William Boyd and Michael Watts, "Agro-Industrial Just-in-Time: The Chicken Industry and Postwar American Capitalism," in Goodman and Watts, eds., *Globalising Food: Agrarian Questions and Global Restructuring* (Routledge, 1997), p. 192–193; "September Is National Chicken Month," *PR Newswire,* Sept. 10, 2001.

2. For an exploration of these questions in the Australian context, see

Jane Dixon, *The Changing Chicken: Chooks, Cooks, and Culinary Culture* (University of New South Wales Press, 2002). Writing for a primarily academic audience, Dixon provides a comprehensive and compelling model for examining a single commodity.

3. John Steele Gordon, "The Chicken Story," *American Heritage* (Sept. 1996). On the overall emergence of the poultry industry, see Gordon Sawyer, *The Agribusiness Poultry Industry: A History of Its Development* (Exposition Press, 1971); John L. Skinner, ed., *American Poultry History, 1823–1973* (American Poultry Historical Society, 1974); William L. Henson, *The U.S. Broiler Industry: Past and Present Status, Practices, and Costs* (Department of Agricultural Economics and Rural Sociology, Agricultural Experiment Station, Pennsylvania State University, 1980); Page Smith and Charles Daniels, *The Chicken Book* (Little, Brown, 1975); and Bernard F. Tobin and Henry B. Arthur, *Dynamics of Adjustment in the Broiler Industry* (Harvard University Graduate School of Business Administration, 1964).

4. Sawyer, *Agribusiness Poultry Industry,* chaps. 1–3. On the famous Mrs. Steele, see William H. Williams, *Delmarva's Chicken Industry: 75 Years of Progress* (Delmarva Poultry Industry, 1998); and Donald D. Stull and Michael J. Broadway, *Slaughterhouse Blues: The Meat and Poultry Industry in North America* (Wadsworth, 2004).

5. Sawyer, *Agribusiness Poultry Industry,* pp. 43–44.

6. Ibid., chap. 3.

7. On the rise of the poultry industry and Tyson in Arkansas see Marvin Schwartz, *Tyson: From Farm to Market* (University of Arkansas Press, 1991); Stephen Strausberg, *From Hills to Hollers: Rise of the Poultry Industry in Arkansas* (Agricultural Experiment Station, University of Arkansas, 1995); W. T. Wilson and R. M. Smith, "Broiler Production and Marketing in Northwestern Arkansas" (Agricultural Experiment Station, University of Arkansas, 1941); and C. Curtis Cable, Jr., "Growth of the Arkansas Broiler Industry" (Agricultural Experiment Station, University of Arkansas, 1952).

8. Interview with Poultry Grower #5, May 28, 2000. I kept the poultry growers I interviewed anonymous (even the ones who are no longer in the business). The interviews are numbered, but my numbering system is not ordered chronologically (so #1 may have been interviewed at a later date than #8). I interviewed forty-eight growers, mostly in North Carolina and Arkansas, but in Georgia and Alabama as well.

9. Interview with Poultry Grower #23, July 28, 2001.

10. Schwartz, *Tyson,* pp. 3–4.

11. Ibid., chap. 1.

12. Sawyer, *Agribusiness Poultry Industry,* chap. 6.

13. Glenn E. Bugos, "Intellectual Property Protection in the American Chicken-Breeding Industry," *Business History Review* 66 (Spring 1992): 127–168, 147.

14. Sawyer, *Agribusiness Poultry Industry,* chap. 6. Also see "An Interview with Jesse Jewell," *Broiler Industry* (Mar. 1959): 8–46.

15. Sawyer, *Agribusiness Poultry Industry,* p. 72.

16. U.S. Congress, Senate Committee on Labor and Public Welfare, J. D. Jewell Co. and Amalgamated Meat Cutters and Butcher Workmen of North America, A.F.L., 82nd Cong., 1st sess., Aug. 9, 1951 (Washington, D.C.: U.S. Government Printing Office, 1951). http://historymatters.gmu.edu/d/6466.html. Thanks to Carl Weinberg for pointing me to this source.

17. Interview with Poultry Grower #14, June 13, 2001.

18. On changes within the poultry industry during World War II, see Sawyer, *Agribusiness Poultry Industry,* chap. 5; Bugos, "Intellectual Property Protection," pp. 127–168; Boyd and Watts, "Agro-Industrial Just-in-Time," pp. 198–199; and Frank Frazier, "Demands of World War II Shaped Poultry Industry," *Feedstuffs,* Aug. 28, 1995, pp. 33–34.

19. Bugos, "Intellectual Property Protection," pp. 147–148.

20. Sawyer, *Agribusiness Poultry Industry,* chap. 5.

21. Boyd and Watts, "Agro-Industrial Just-in-Time," pp. 197–200; Tobin and Arthur, *Dynamics of Adjustment;* and Bugos, "Intellectual Property Protection," pp. 145–155.

22. Bugos, "Intellectual Property Protection," p. 148.

23. William Boyd, "Making Meat: Science, Technology, and American Poultry Production," *Technology and Culture* 42 (Oct. 2001): 631–664.

24. Ibid., pp. 638, 636–637.

25. For a detailed, yet brief and readable, description of how growers raise a modern chicken, see Stull and Broadway, *Slaughterhouse Blues,* pp. 43–48.

26. Boyd and Watts, "Agro-Industrial Just-in-Time," p. 199.

27. Ibid., p. 200.

28. Ibid.

29. Ibid., pp. 211–212.

30. Ibid., p. 212.

31. Interview with Poultry Grower #29, July 15, 2001.

32. On the varied nature of growers' opinions, see Stull and Broadway, *Slaughterhouse Blues.*

33. Interview with Poultry Grower #23, June 24, 2001.

34. There is a rich literature on poultry growers. Steve Bjerklie has a helpful statement of the issues involved in "Dark Passages," *Meat and Poul-*

try, parts 1 and 2 (Aug./Oct. 1994). For an interesting longitudinal study, see William D. Heffernan and David H. Lind, "Changing Structure in the Broiler Industry: The Third Phase of a Thirty-Year Longitudinal Study (Unpublished report available from the authors at the Department of Rural Sociology, University of Missouri, Columbia, 2000). Stull and Broadway, in their *Slaughterhouse Blues,* provide an excellent case study for a poultry region in Kentucky.

35. B. W. Marion and H. B. Arthur, "Dynamic Factors in Vertical Commodity Systems: A Case Study of the Broiler System" (Ohio Agricultural Research and Development Center, Wooster, 1973).

Chapter 3
Anatomy of a Merger

Epigraph: Interview with Poultry Grower #19, July 8, 2001.

1. Interview with Poultry Grower #21, July 22, 2001.

2. There is no scholarly history of Holly Farms. There are various accounts of the company's history in newspapers, both local and national, as well as in industry magazines and company publications. For a brief historical overview of the company, see "A Company That Started from Scratch," *Holly Happenings* 15, no. 5 (Apr. 1986). See also "How Holly Fits with Tyson," *Tyson Updates* (Oct. 1989): 4–8.

3. "Fred Lovette: His Industry Helped Set Wilkes on the Highway to Better Times," *Winston-Salem Journal,* Nov. 1, 1987; "The Lovettes Really Started Something: Small Farm Route in Wilkes Has Become Million-Dollar Industry," *Winston-Salem Journal,* Nov. 6, 1955.

4. "Farm Forum: Wilkesboro Poultry Plant Puts Up Chickens Ready for the Pan," *Winston-Salem Journal,* June 27, 1948.

5. Ibid.

6. "Fred Lovette."

7. Ibid.; "A Company Started from Scratch," *Wilkes Journal-Patriot,* Mar. 28, 1997.

8. "A Company Started from Scratch."

9. "Holly Farms Poultry," *Arbor Acres Review* 5, no. 8 (Apr. 1962); "1961 Merger to Form Holly Farms Was Done for Economic Survival," *Wilkes Journal-Patriot,* Apr. 27, 1995.

10. "Holly Farms Poultry"; "1961 Merger."

11. For a detailed discussion of the innovative nature of Holly Pak see "Holly Revisited," *Broiler Industry* (Apr. 1968): 23–47.

12. North Carolina Citizens Association, "A Name Housewives Recognize in the Poultry Display Case," *We the People of North Carolina* (official publication of the North Carolina Citizens Association, Raleigh), Dec. 1979.

13. Ibid.

14. Holly Farms annual reports, various years; "Holly Farms CEO Started at Chicken Coop," *Charlotte Observer*, Oct. 19, 1987.

15. "Holly Is to Buy Pa. Poultry Firm," *Wilkes Journal-Patriot*, Feb. 1, 1988.

16. "Problems with Roast Chicken," *Wilkes Journal-Patriot*, Feb. 15, 1988.

17. William Glaberson, "Holly Expected to Fight Tyson Offer," *New York Times*, Oct. 13, 1988, p. D1.

18. Associated Press, "Tyson Foods Bids for Holly Farms," *New York Times*, Oct. 12, 1988, p. D1.

19. Financial Desk, "Holly Farms Rejects Tyson Bid," *New York Times*, Oct. 19, 1988, p. D4.

20. Glaberson, "Holly Expected to Fight."

21. Guy Halverson, "Rush Is On to Get Takeovers Wrapped Up," *Christian Science Monitor*, Oct. 24, 1988, p. 19.

22. Nina Andrews, "Holly Will Consider New Tyson Foods Bid," *New York Times*, Oct. 22, 1988, sec. 1, p. 35; Robert Vincent, "Tyson Foods Sweetens Offer for Holly," *Financial Times*, Oct. 22, 1988, sec. 1, p. 12.

23. Nina Andrews, "Tyson Foods Raises Bid for Holly Farms," *New York Times*, Dec. 1, 1988, p. D5; Reuters, "Tyson Bid Rejected by Holly Farms," *New York Times*, Dec. 8, 1988, p. D4; Associated Press, "Tyson Foods–Holly," *New York Times*, Dec. 14, 1988, p. D5.

24. Associated Press, "Tyson Wins Holly Round," *New York Times*, Dec. 31, 1988, sec. 1, p. 33; "Holly's Board Sets Auction," *New York Times*, Jan. 10, 1989, p. D4; Associated Press, "Holly Farms to Test Ruling," *New York Times*, Jan. 4, 1989, p. D5.

25. "Tyson Sweetens Holly Farms Bid," *New York Times*, Jan. 20, 1989, p. D4; Deborah Hargreaves, "Tyson Set to Devour Holly Farms," *Financial Times*, Jan. 10, 1989, sec. 1, p. 23.

26. "Holly Board Backs ConAgra," *New York Times*, Feb. 8, 1989, p. D4; Nina Andrews, "Holly Farms Holders Veto ConAgra Bid," *New York Times*, Apr. 18, 1989, p. D5.

27. "Holly Farms Chicken OKs ConAgra Merger," *St. Louis Post-Dispatch*, May 22, 1989, p. 8A; Michael Freitag, "New Deal by ConAgra and Holly," *New York Times*, May 22, 1989, p. D1; Paul Wiseman, "LIN spurns McCaw," *USA Today*, June 21, 1989, p. 3B; Glenn Ruffenach, "Holly Farms Battle May Be Nearing an End," *Wall Street Journal*, June 23, 1989, sec. 1, p. 3; Financial Desk, "Holly in Talks to Call Off Deal," *New York Times*, June 23,

1989, p. D3; Reuters, "Holly Accepts Tyson Offer Ending Lengthy Chicken War," *Toronto Star*, June 24, 1989, p. C4.

28. Nina Andrews, "Tyson Bid Is Accepted by Holly," *New York Times*, June 24, 1989, sec. 1, p. 31.

29. "Chicken King, Under Siege, Looked into His Mirror," *St. Louis Post-Dispatch*, July 9, 1989, p. 5E.

30. Anonymous interview, Summer 2001.

31. John Taylor, "Blake Lovette Will Head ConAgra Unit," *Omaha World Herald*, Sept. 13, 1989, p. 3m; "Economy Called More Stable: Sales Are Up," *Wilkes Journal-Patriot*, Oct. 12, 1989.

Chapter 4
The Right to Work

1. "Tyson Reassures Holly People," *Wilkes Journal-Patriot*, July 20, 1989; "Tyson Says No Job Losses," *Wilkes Journal-Patriot*, June 26, 1989.

2. "Restructuring Seen as Part of Movement to Arkansas," *Wilkes Journal-Patriot*, Aug. 9, 1990.

3. "Tyson Is Continuing to Cut Wilkes Jobs," *Wilkes Journal-Patriot*, Oct. 11, 1990; Jule Hubbard, "Sale of Tyson Building Eyed," *Wilkes Journal-Patriot*, Feb. 7, 1991; "Tyson Is Moving Jobs to Springdale," *Wilkes Journal-Patriot*, June 3, 1991; "Don Tyson Sees Bright Future in Wilkes," *Wilkes Journal-Patriot*, Apr. 30, 1989.

4. Interview with Trucker #4, July 26, 2001. To complement the newspaper accounts and court documents, I interviewed five of the truckers involved in the dispute.

5. "Vote 'No' to Union," editorial, *Wilkes Journal-Patriot*, Jan. 26, 1989; "Teamsters' Big Win Could Spread Union," *Wilkes Journal-Patriot*, Mar. 23, 1989.

6. Jule Hubbard, "More Workers at Holly Farms Seek a Union," *Wilkes Journal-Patriot*, Feb. 29, 1989. Agricultural workers have been historically left out of major labor legislation, particularly the National Labor Relations Act of 1935, which effectively required nonagricultural employers to recognize and bargain with labor unions.

7. "Labor Hearing Delayed Due to Death," *Wilkes Journal-Patriot*, Jan. 8, 1990.

8. Ibid.; "Teamsters Vacate Picket Line Outside Tyson," *Wilkes Journal-Patriot*, Jan. 20, 1992.

9. Jule Hubbard, "In ESC Ruling Drivers Favored," *Wilkes Journal-Patriot*, Mar. 22, 1990.

10. "Tyson Won't Immediately Rehire," *Wilkes Journal-Patriot,* Mar. 23, 1992.

11. Teamsters Local 391, press release, Apr. 23, 1996.

12. Jule Hubbard, "Labor Ruling Would Cost Tyson Millions," *Wilkes Journal-Patriot,* Mar. 16, 1995.

13. "Tyson Is Cutting Costs," *Wilkes Journal-Patriot,* Jan. 11, 1996. A couple of years later, in March 1998, the workers scored another, even more important, victory. Holly first, then Tyson, had argued that live production workers, particularly chicken catchers and haulers, were agricultural workers and could not unionize. The Supreme Court disagreed, ruling that chicken catchers, live-haul drivers, and forklift operators are employees and not agricultural workers. They could indeed unionize.

14. Interview with Trucker #2, July 12, 2001.

15. Interview with Poultry Grower #33, Aug. 1, 2001. I interviewed approximately twenty of the Mountain Growers.

16. Superior Court Division, General Court of Justice, [Early 1990s] file 90, CVS 98, Ashe County, North Carolina.

17. Interview, July 18, 2001.

18. Ibid.

19. Superior Court Division, file 90, CVS 98.

20. Ibid.; "Chicken Economics: The Broiler Business Consolidates, and That Is Bad News to Farmers," *Wall Street Journal,* Jan. 4, 1990, p. A1; Interview, July 18, 2001.

21. Interview, July 18, 2001.

22. "Chicken Economics."

23. Interview, July 18, 2001.

24. Ibid.

25. Superior Court Division, file 90, CVS 98.

26. Interview with Poultry Grower #36, July 7, 2001.

27. "Tyson Reassures Holly People."

28. Interview with Poultry Grower #37, July 8, 2001.

29. Interview with Poultry Grower #38, July 11, 2001.

30. Interview with Joe Brown, July 9, 2001.

31. "Ashe and Watauga Farms May Have Chickens Again," *Wilkes Journal-Patriot,* Feb. 21, 1994.

32. Interview with Poultry Grower #34, July 9, 2001.

33. Ibid.

34. "Gold Kist Official Comments on Assistance," *Wilkes Journal-Patriot,* Mar. 27, 1995; Interview with Poultry Grower #40, July 20, 2001.

35. Interview with Poultry Worker #37, July 13, 2001.

Chapter 5
Getting Here

1. Kevin Sack, "Under the Counter, Grocer Provided Immigrant Workers," *New York Times,* Jan. 14, 2002, www.nytimes.com.

2. Ibid.

3. Ibid.

4. Manuel Torres, "The Latinization of the South," *Mobile Register,* June 28, 1999, p. 1A; Russell Cobb, "The Chicken Hangers," *Identify,* Feb. 2, 2004.

5. U.S. Census 1990, 2000.

6. Raymond Mohl, "Globalization, Latinization, and the Nuevo New South," *Journal of American Ethnic History* 22, no. 4 (2003): 31–66; William Kandel and Emilio A Parrado, "Industrial Transformation and Hispanics in the American South: The Case of the Poultry Industry," in *Hispanic Spaces, Latino Places: A Geography of Regional and Cultural Diversity,* ed. Daniel D. Arreola (University of Texas Press, 2004); William Kandel and Emilio Parrado, "U.S. Meat-Processing Industry Restructuring and New Hispanic Migration," paper presented at the 2003 Annual Meeting of the Population Association of America, Minneapolis, May 1–3, 2003.

7. For an excellent collection of case studies describing the entrance, reception, and experience of Latin-American immigrants in the South, see Arthur D. Murphy et al., *Latino Workers in the Contemporary South* (University of Georgia Press, 2001); statistics are from Mohl, "Globalization," pp. 33, 44.

8. On the reception of immigrants into poultry producing regions, see Greig Guthey, "Mexican Places in Southern Spaces: Globalization, Work, and Daily Life in and Around North Georgia," in Murphy et al., *Latino Workers in the Contemporary South*; and Leon Fink, *The Maya of Morganton: Work and Community in the Nuevo New South* (University of North Carolina Press, 2003). One of the best discussions of this process, and explanations of how communities respond to new arrivals, can be found in Donald D. Stull and Michael J. Broadway, *Slaughterhouse Blues: The Meat and Poultry Industry in North America* (Wadsworth, 2004), chaps. 7–9. For an interesting discussion of "labor shortages" within the poultry industry, see David Griffith, *Jones's Minimal: Low-Wage Labor in the United States* (State University of New York Press, 1993).

9. On recent changes brought about by meatpacking, see Donald D. Stull, et al., eds., *Any Way You Cut It: Meat Processing and Small-Town America* (University Press of Kansas, 1995); Stull and Broadway, *Slaughterhouse Blues;* Deborah Fink, *Cutting into the Meatpacking Line: Workers and Change in the Rural Midwest* (University of North Carolina Press, 1998); Louise

Lamphere et al., eds., *Newcomers in the Workplace: Immigrants and the Restructuring of the U.S. Economy* (Temple University Press, 1994).

10. Statistics don't tell the whole story, but they are revealing. The correlation between poultry producing regions and Hispanic immigration is dramatic, particularly in America's heartland. In the South the trends are even more obvious. Eight of the nine poultry producing states in the South are among the ten states with the fastest growing "nonmetropolitan" Hispanic populations, and the next ten states on the list are Midwestern and have large beef- and pork-processing industries. Narrow our gaze a bit more, and focus on counties, and the correlation between poultry processing and Latin-American immigration is even more transparent. See Kandel and Parrado, "Industrial Transformation and Hispanics," pp. 2–5; Kandel and Parrado, "U.S. Meat Processing."

11. Sack, "Under the Counter."

12. Ibid.

13. Ethan Nobles, "Analyst Looks for Tyson to Weather Indictment," *Morning News of Northwest Arkansas* (Springdale), Dec. 20, 2001.

14. Barry Yeoman, "Hispanic Diaspora," *Mother Jones* 25, no. 4 (July/Aug. 2000): 34.

15. Kyle Mooty, "Tyson Foods Faces Charges of Smuggling Illegal Aliens," *Arkansas Business*, Dec. 19, 2001, www.arkansasbusiness.com.

16. Cristal Cody, "Indictment Claims Tyson Conspired to Smuggle Illegal Aliens to Employ," *Arkansas Democrat Gazette*, Dec. 20, 2001, nwa.com.

17. Greg Farrell, "Tyson Recruited Illegals, INS Says," *USA Today*, Dec. 20, 2001.

18. These figures are notoriously difficult to calculate, in part because fake documents are relatively easy to acquire. Percentages of course vary depending on region and even plant. But somewhere between one-quarter and one-third is considered likely.

19. Sherri Day, "Jury Clears Tyson Foods in Use of Illegal Immigrants," *New York Times*, Mar. 27, 2003.

20. Bill Poovey, "Tyson Indictments Leave Some Illegal Immigrants Stranded," Associated Press (2003), www.imdiversity.com/villages/hispanic/Article_ Detail.asp?Article_ID=9053.

21. As David Griffith points out, many companies encourage workers (often with financial incentives) to recruit their family and friends. See Griffith, *Jones's Minimal*, pp. 159–163. For an informative account of worker recruitment and travel to poultry producing regions, see Jesse Katz, "The Chicken Trail," *Los Angeles Times*, Nov. 10–12, 1996.

22. Interview, June 13, 2001.

23. Interview, June 12, 2001. There are many accounts of the migration process. For one of the richest and most complex, see Ruben Martinez, *Crossing Over: A Mexican Family on the Migrant Trail* (Picador, 2001).

24. Interview, Dec. 28, 2000.

25. Interview, Nov. 13, 2001.

26. Interview with Poultry Worker #31, Dec. 3, 2001.

27. Robert Robb, "Tightened U.S. Border Policy Keeps Mexicans Trapped," *Tucson Citizen*, Dec. 3, 2002.

28. Esteban's story was documented by Russell Cobb in his article "Chicken Hangers."

29. Ibid.

30. Ibid.

31. Ibid.

Chapter 6
Inside a Poultry Plant

1. For a broader understanding of the gender and ethnic composition of plant labor forces, see David Griffith, *Jones's Minimal: Low-Wage Labor in the United States* (State University of New York Press, 1993).

2. It was a poultry worker who suggested I get a job in a processing plant. "If you want to understand chicken, you need to work in a plant." When applying for the job, I did not reveal to Tyson Foods that I was an anthropology professor. Such a practice, although fairly common within investigative journalism, is not without controversy within the field of anthropology, though it is not unheard of; see Deborah Fink, *Cutting into the Meatpacking Line: Workers and Change in the Rural Midwest* (University of North Carolina Press, 1998). I decided to not tell Tyson for two basic reasons. First, I probably would not have been allowed in the plant, and I strongly believe that these are the kinds of field sites anthropologists need to be investigating. Also, there was virtually no chance that anyone would be harmed by my presence. Second, as is often the case with anthropological participant-observation, I felt it was important that everyone behave normally. That is, I did not want to be treated differently; nor did I want workers or managers to act differently—as they might if they knew I was an anthropologist. I did eventually tell some coworkers that I was an anthropologist. They all agreed that one had to work in a plant in order to understand the poultry industry and tell this important story.

3. Although much of the South was going through something of a boom during this period, northwest Arkansas was doing particularly well

(which meant that white workers got out of plants more quickly and decisively than in other regions). In short, the timing and intensity of the departure of U.S.-born workers and the arrival of a foreign labor force varied from region to region (and often from plant to plant).

4. Interview, Oct. 11, 2000.

5. Interview, Oct. 22, 2000.

6. Interview, Aug. 10, 2001.

7. Measuring the relative danger of occupations is tricky not only because what is being measured can vary (fatal versus nonfatal injuries, injury versus illness, etc.), but also because such statistics depend on companies to keep accurate records—a serious problem within the meat industries. Statistics on injuries can also depend on whether a worker missed work, which means injured workers who continue to work are often not counted. Regardless, meatpacking is routinely among those industries with the highest rates of workplace injury (see the Web site for the U.S. Dept. of Labor, Bureau of Labor Statistics: www.bls.gov). The poultry industry is less dangerous than the beef industry, but for much of its history it has suffered twice the average private-sector rates of illness and injury. Within the poultry industry as a whole, injuries are generally associated with the raising of live birds, which entails problems associated with respiration, or the slaughtering, processing, and packaging of chickens (which includes not only ergonomic problems, but cuts, trips and falls, electrical injuries, etc.). For a compelling (if somewhat dated) discussion of the danger associated with the poultry workplace, see Barbara Goldoftas, "To Make a Tender Chicken: Poultry Workers Pay the Price," *Dollars and Sense: The Magazine of Economic Justice* 242 (July/Aug. 2002 [1989]).

8. Human Rights Watch, *Blood, Sweat, and Fear: Workers' Rights in U.S. Meat and Poultry Plants* (Human Rights Watch, 2005), p. 24.

9. Interview with Poultry Worker #13, Aug. 6, 2001.

10. Interview with Poultry Worker #3, Sept. 20, 2000.

11. Interview with Poultry Worker #8, Nov. 1, 2000.

Chapter 7
Growing Pains

1. Jeff South and David Kennamer, "Adapting to Rapid Change," *Quill* 91, no. 2 (Mar. 2003): 30; Eric Schmitt, "Pockets of Protest Are Rising Against Immigration," *New York Times*, Aug. 9, 2001, p. A12; Linda Robinson, "Hispanics Don't Exist," *U.S. News and World Report*, May 11, 1998, p. 26.

2. Barry Yeoman, "Hispanic Diaspora," *Mother Jones* 25, no. 4 (July/

Aug. 2000): 34; Bob Cook, "Old Town, Nuevo Patients," *American Medical News,* Oct. 1, 2001, p. 26.

3. Raymond Mohl, "Latinization in the Heart of Dixie: Hispanics in Late-Twentieth Century Alabama," *Alabama Review* 55 (Oct. 2002). Immigrants entered the poultry industry in some parts of the South during the 1980s, but the real surge occurred during the 1990s. See David Griffith, *Jones's Minimal: Low-Wage Labor in the United States* (State University of New York Press, 1993); South and Kennamer, "Adapting to Rapid Change," p. 30; Cook, "Old Town, Nuevo Patients," p. 26.

4. John D. Studstill and Laura Nieto-Studstill, "Hospitality and Hostility: Latin Immigrants in Southern Georgia," in Arthur Murphy et al., eds., *Latino Workers in the Contemporary South* (University of Georgia Press, 2001); Mohl, "Latinization in the Heart of Dixie."

5. Michael Barone et al., "The Many Faces of America," *U.S. News and World Report,* Mar. 19, 2001, p. 18.

6. There is a growing literature on this topic. For the South, see Mohl, "Latinization in the Heart of Dixie"; Studstill and Nieto-Studstill, "Hospitality and Hostility"; and Greig Guthey, "Mexican Places in Southern Places: Globalization, Work, and Daily Life in and Around the North Georgia Poultry Industry," in Arthur D. Murphy, et al., *Latino Workers in the Contemporary South* (University of Georgia Press, 2001). For meatpacking and the Midwest, see Donald D. Stull and Michael J. Broadway, *Slaughterhouse Blues: The Meat and Poultry Industry in North America* (Wadsworth, 2004); and Donald D. Stull et al., eds., *Any Way You Cut It: Meat Processing and Small-Town America* (University Press of Kansas, 1995). For one of the best case studies of immigrant–native relations, see the special issue on Garden City in *Urban Anthropology* 19, no. 4 (1990).

7. Anne Krishnan, "Businesses Learn Value of Marketing to Hispanic Customers in Durham, N.C.," *Knight Ridder Tribune Business News,* Sept. 14, 2003, p. 1.

8. For reasons that will become apparent in this chapter, Siler City has received considerable media attention. This section relies on those accounts, especially the wonderful reporting of Barry Yeoman, as well as several trips that I took to Siler City and interviews with fourteen residents. See Yeoman, "Hispanic Diaspora"; and David Bailey, "The Right Place to Bee," *Business, North Carolina* 14, no. 4 (1994): sec. 1, p. 50.

9. Yeoman, "Hispanic Diaspora."

10. Ibid.

11. Bailey, "The Right Place to Bee," p. 50; Jennifer Toth, "Meanwhile, in the Other South," *Business Week,* Sept. 27, 1993, p. 104.

12. South and Kennamer, "Adapting to Rapid Change," p. 30.

13. Ibid.; Sue Anne Pressley, "Hispanic Immigration Boom Rattles South," *Washington Post,* Mar. 6, 2000, p. A3; Pressley, "Hispanic Immigration Boom," A3.

14. Yeoman, "Hispanic Diaspora."

15. Ibid.

16. Interview with Poultry Worker #5, Mar. 3, 2000.

17. Interview with Arkansan #3, Feb. 18, 2002.

18. Interview with Arkansan #8, Sept. 12, 2002; Interview with North Carolinian #2, Oct. 22, 2000; Pressley, "Hispanic Immigration Boom," p. A3.

19. Yeoman, "Hispanic Diaspora."

20. Schmitt, "Pockets of Protest," p. A12; Sergio Bustos, "Small Towns Shaped by Influx of Hispanics," *USA Today,* May 23, 2000, p. A10.

21. Yeoman, "Hispanic Diaspora"; Russell Cobb, "The Chicken Hangers," *Identify,* Feb. 2, 2004.

22. Manuel Torres, "The Latinization of the South," *Mobile Register,* June 28, 1999, p. A4.

23. Bustos, "Small Towns Shaped," p. A10.

24. Schmitt, "Pockets of Protest," p. A12.

25. For an interesting case of racial tension on the plant floor, see Charlie LeDuff, "At a Slaughterhouse, Some Things Never Die," *New York Times,* June 16, 2000.

26. Interview with Arkansan #12, Oct. 1, 2002.

27. Yeoman, "Hispanic Diaspora."

28. Ibid.

29. Ibid.

30. Ibid.

31. Ibid.

32. Teresa Malcolm, "Church Prays in Face of Rally Against Immigrants," *National Catholic Reporter,* Mar. 3, 2000, p. 8.

33. Pressley, "Hispanic Immigration Boom," p. A3; Yeoman, "Hispanic Diaspora."

34. Pressley, "Hispanic Immigration Boom," p. A3.

35. Leda Hartman, "Significant Latino Population Growth in Siler City, North Carolina," *National Public Radio* (June 7, 2001).

36. Interview with North Carolinian #10, May 20, 2001.

37. Interview with Poultry Worker #29, May 18, 2001.

38. Hartman, "Significant Latino Population Growth."

39. Eric Schmitt, "Whites in Minority in Largest Cities, the Census Shows," *New York Times,* Apr. 30, 2001, p. A1.

40. For an interesting example, see Studstill and Nieto-Studstill, "Hospitality and Hostility."

41. Stull and Broadway, *Slaughterhouse Blues,* p. 102.

42. Torres, "Latinization of the South," p. 4A.

Chapter 8
Toward a Friendlier Chicken

Epigraph: Steve Lustgarden and Debra Holton, "What About Chicken?" *EarthSave: Healthy People, Healthy Planet* (1997). www.earthsave.org/news/chicken.htm.

1. www.tyson.com. Viewed Apr. 2004.

2. For a good discussion of CAFOs and the surrounding controversy, see Donald D. Stull and Michael J. Broadway, *Slaughterhouse Blues: The Meat and Poultry Industry in North America* (Wadsworth, 2004).

3. Although the Sierra Club has brought the cases together on its Web site, where they will be easy for readers to review, I also consulted the original sources: government regulating such as the Environmental Protection Agency (EPA), Department of Labor, Occupational Safety and Health Administration (OSHA), and the Centers for Disease Control and Prevention (CDC).

4. Human Rights Watch, *Blood, Sweat, and Fear: Workers' Rights in U.S. Meat and Poultry Plants* (Human Rights Watch, 2005), pp. 1–2.

5. Richard Behar and Michael Kramer, "Something Smells Fowl," *Time,* Oct. 17, 1994; Lustgarden and Holton, "What About Chicken?"

6. Lustgarden and Holton, "What About Chicken?"

7. Ibid.

8. Ibid. This article nicely summarizes two studies from the Government Accountability Project.

9. Europeans have a more highly developed system and market for alternative forms of poultry. One of the most interesting is Label Rouge in France, which raises poultry on pasture and then markets premium chicken using its own brand. See Anne Fanatico and Holly Born, "Label Rouge: Pasture Based Poultry Production in France," Livestock Technical Note (2002). http://www.attra.org/attra-pub/labelrouge.html. Thanks to Anne for educating me about this movement.

10. For a wonderful discussion of the potential benefits of organic food (and contradictions in the industry), see Michael Pollan, "Naturally, How Organic Became a Marketing Niche and a Multimillion Dollar Indus-

try," *New York Times Magazine,* May 13, 2001. For more information, see Michael Shuman, *Bay Friendly Chicken: Reinventing the Delmarva Poultry Industry,* study sponsored by the Chesapeake Bay Foundation and the Delmarva Poultry Justice Alliance (2000). The company's Web site (www.bay friendlychicken.com) should be up by the time this book is in print. I thank Michael Shuman for talking with me about Bay Friendly Chicken.

Index

The Agrarian Studies Series at Yale University Press seeks to publish outstanding and original interdisciplinary work on agriculture and rural society—for any period, in any location. Works of daring that question existing paradigms and fill abstract categories with the lived-experience of rural people are especially encouraged.

JAMES SCOTT, *Series Editor*

Christiana Payne, *Toil and Plenty: Images of the Agricultural Landscape in England, 1780–1890* (1993)

Brian Donahue, *Reclaiming the Commons: Community Farms and Forests in a New England Town* (1999)

James Scott, *Seeing Like a State: How Certain Schemes to Improve the Human Condition Have Failed* (1999)

Tamara L. Whited, *Forests and Peasant Politics in Modern France* (2000)

Nina Bhatt and James C. Scott, *Agrarian Studies: Synthetic Work at the Cutting Edge* (2001)

Peter Boomgaard, *Frontiers of Fear: Tigers and People in the Malay World, 1600–1950* (2001)

Janet Vorwald Dohner, *The Encyclopedia of Historic and Endangered Livestock and Poultry Breeds* (2002)

Deborah Fitzgerald, *Every Farm a Factory: The Industrial Ideal in American Agriculture* (2003)

Stephen B. Brush, *Farmer's Bounty: Locating Crop Diversity in the Contemporary World* (2004)

Brian Donahue, *The Great Meadow: Farmers and the Land in Colonial Concord* (2004)

J. Gary Taylor and Patricia J. Scharlin, *Smart Alliance: How a Global Corporation and Environmental Activists Transformed a Tarnished Brand* (2004)

Raymond L. Bryant, *Nongovernmental Organizations in Environmental Struggles: Politics and the Making of Moral Capital in the Philippines* (2005)

Edward Friedman, Paul G. Pickowicz, and Mark Selden, *Revolution, Resistance, and Reform in Village China* (2005)

Michael Goldman, *Imperial Nature: The World Bank and Struggles for Social Justice in the Age of Globalization* (2005)

Arvid Nelson, *Cold War Ecology: Forests, Farms, and People in the East German Landscape, 1945–1989* (2005)

Steve Striffler, *Chicken: The Dangerous Transformation of America's Favorite Food* (2005)